# MICROBIOLOGY AND BIOTECHNOLOGY
## A Laboratory Manual

MJP
PUBLISHERS

# MICROBIOLOGY AND BIOTECHNOLOGY
## A Laboratory Manual

*P.T. Kalaichelvan*
Professor
CAS in Botany
University of Madras
Chennai

Chennai　　　　Trichy　　　　New Delhi

ISBN 978-81-8094-008-8 **MJP PUBLISHERS**

All rights reserved No.44, Nallathambi Street
Triplicane, Chennai 600 005

MJP 007 © Publishers, 2019

Publisher: C.Janarthanan

*To*

*my wife Akila*

*&*

*daughter Aparanjini*

# FOREWORD

Microbiology and biotechnology are the two important practical-oriented subjects in the biology curriculum. Observation, study and experimentation are the basic needs for these subjects. Prof. P. T. Kalaichelvan, based on his experience of over two decades has made an excellent effort to bring out this volume on basic experiments that are required to be carried out by all students who study these subjects. Experiments in microbiology include counting of microbes, identification of bacteria, culturing and isolation of microbes, enzyme production and immobilization techniques. Isolation of genomic DNA of prokaryotes and eukaryotes, isolation of plasmids, transformation in E.coli, isolation of bacteriophages, ELISA technique, electrophoretic techniques, Western blotting protocol, preparation of protoplasts and micropropagation are the experiments for biotechnology. This book will serve as a useful manual for students and teachers in these fields and I congratulate Prof. Kalaichelvan for taking efforts to bring out this book.

*Prof. N. Anand, D.Sc.*
Director, CAS in Botany
University of Madras
Chennai

# PREFACE

I am extremely happy to bring out this practical guide on experiments in the fields of microbiology and biotechnology. I am confident that it would serve the purpose of being useful to students as a dependable laboratory manual.

This book covers basic topics like enumeration of microbes as well as advanced techniques like isolation of genomic DNA and transformation. The techniques that I have standardized have been illustrated in this book. The chemical composition of different media and their preparation are presented in the appendices.

I am grateful to Prof. N. Anand, Director, CAS in Botany, University of Madras for his constant guidance in this endeavour. I also thank my dear friend Prof. R. Rengasamy for his valuable suggestions. I am indebted to Mr. Pillai of MJP Publishers for bringing out this book within a short time.

*P.T.Kalaichelvan*

# CONTENTS


Safety Guidelines     xiii

Microbial Cell Counting     1

   1. Measurement of Cells of Bacteria or Fungal Conidia
       by a Counting Chamber     3
   2. Enumeration of Bacteria by Serial Dilution Agar
       Plate Technique     9

Microscopic Observation of Microorganisms     15

   3. Wet Mount     17
   4. Hanging Drop Technique for Demonstrating
       Motility of Bacteria     21
   5. Negative Stain     25
   6. Simple Stain     31
   7. Gram Stain     37
   8. Acid-fast Stain     43
   9. Spore Stain     49
 10. Capsule Stain     55
 11. Slide Culture Technique     61

 12. Preparation and Inoculation of Growth Media     67
 13. Culture Characterisation of Bacteria     79
 14. Biochemical Tests Used to Identify Bacteria     83
 15. Assessing Antibiotic Effectiveness: The Kirby–Bauer Method     91
 16. Screening of Amylase-Producing Organisms     99
 17. Screening of Cellulase-Producing Microorganisms     103
 18. Immobilised Whole Cells     107
 19. Making Sauerkraut     113
 20. Making Ginger Beer     115
 21. Ethanoic Acid Production     119
 22. Isolation of Genomic DNA     123
 23. Isolation of Plasmid DNA     125

24.  Rapid Isolation of Plasmid using TELT                                    129
25.  Visualisation of DNA in Agarose Gel                                     131
26.  Quantitation of DNA by Spectrophotometry                               135
27.  Restriction Enzyme Digestion and Size Determination
     of DNA                                                                 139
28.  Transformation in *E. coli*                                            143
29.  NTG–Mutagenesis of *E. coli*                                           147
30.  Ampicillin Selection of Auxotrophs                                     151
31.  Mutagenesis in Bacteria: The Ames Test                                 155
32.  Isolation of Bacteriophage from Sewage and Determination
     of Phage Titer                                                         161
33.  Identification and Enumeration of White Blood Cells                    169
34.  Antigen–Antibody Precipitation Reactions and
     Determination of Antibody Titer                                        177
35.  Agglutination Reactions: ABO Blood Typing                              183
36.  Immunodiffusion: Antigen–Antibody Precipitation
     Reactions in Gels                                                      189
37.  Enzyme-linked Immunosorbent Assay (ELISA)                              199
38.  SDS–Polyacrylamide Gel Electrophoresis                                 207
39.  Western Blotting                                                       213
40.  Measurement of Plant Cell Growth                                       219
41.  Establishing a Plant Culture                                           223
42.  Micropropagation of African Violet                                     227
43.  Preparation and Fusion of Protoplasts                                  231
44.  Isolation of Chloroplasts from Spinach Leaves                          235
45.  Isolation of plant DNA                                                 239
     Appendix I                                                             243
     Appendix II                                                            249

# SAFETY GUIDELINES

## GENERAL GUIDELINES FOR EVERY LAB SESSION

1. Wear appropriate clothing and shoes to the laboratory. Shoes must completely cover the feet to provide protection from broken glass and spills.
2. Place all books, backpacks, purses, etc. in an area designated by your laboratory instructor. Carry to your work area, only the items you will use in the lab.
3. Wash your work area with disinfectant, and allow air-drying before beginning the lab session.
4. Wipe your work area with disinfectants, and allow air-drying before beginning the lab session.
5. Do not perform activities in the lab until your laboratory instructor gives you instructions.
6. Do not eat, drink, smoke, or apply make-up while working in the laboratory.
7. If you happen to cut or burn yourself while working, report this immediately to your laboratory instructor.
8. Broken glassware should be immediately brought to the attention of your laboratory instructor; broken glass should be placed in a special sharps container for disposal and not in the trash container.
9. While using a Bunsen burner, tie back long hair. Do not lean over the countertop. When in use, always be aware of the flame. Keep flammable items away from the flame. Turn off the burner when not in use.
10. Before leaving the lab, make sure all items have been returned to their appropriate location.
11. After your work is clear, wipe down your countertop with disinfectant before leaving.
12. Wash your hands thoroughly with antibacterial soap before leaving the lab.
13. Do not remove any item from the lab unless you have been directed to do so by the laboratory instructor.

## GUIDELINES FOR WORKING WITH BACTERIA

Handling live bacteria in the laboratory, even those considered nonpathogenic, requires special guidelines beyond the general guidelines already mentioned. All bacteria are potentially pathogenic, especially if they gain entry into the human body. So observe the following guidelines when handling nonpathogenic bacteria listed.

1. Do not put anything into your mouth when working with cultures. Do not pipette by mouth; use a pipette aid instead. Keep your hands, pencil, pen, etc. away from your mouth, eyes, and nose.
2. When inoculating cultures, sterilise the loop or needle before placing it on the counter.
3. Always keep tubes in test tube racks when working with liquid media. Do not stand them up or lay them down on the countertop where they may spill.
4. If you accidentally spill a culture, cover the spill with a paper towel, flood it with disinfectant, and inform your laboratory instructor.
5. Place all used culture media, paper towels, gloves, etc. into the waste container designated by your laboratory instructor. A separate waste container for sharps (slides, pipettes, swabs, broken glass, etc.) will also be provided. All this waste will be autoclaved before disposal or reuse. Do not throw any of these items into the trash container.
6. If you have a burn or wound on one of your hands, cover it with a plastic strip and wear disposable gloves for added protection.

Handling pathogenic bacteria in the laboratory requires special guidelines beyond the general guidelines and those for nonpathogenic bacteria. The following additional guidelines apply when working with pathogenic bacteria.

1. When handling pathogens, access to the laboratory must be restricted to only those working in the lab.
2. Disposable gloves and a lab coat must be worn. The gloves should be disposed off in a container designated by the instructor. The lab coat must be removed before leaving and kept in a designated area of the lab.
3. Avoid creating aerosols when working with pathogens, and if there is a chance of creating tiny airborne droplets, work under a safety hood.

## UNIVERSAL PRECAUTIONS

All human blood and certain other body fluids are treated as if they are infectious for blood-borne pathogens, such as human immuno-deficiency virus (HIV), hepatitis B virus (HBV), and hepatitis C virus (HCV). Such precautions are the rule among nurses, doctors, phlebotomists, and clinical laboratory personnel, and are a critical component of infection control.

1. Wear gloves.
2. Change gloves when they are soiled or torn.
3. Remove gloves when you are finished handling a specimen, and before you touch other objects such as drawer handles, doorknobs, refrigerator handles, pens/pencils, and paper.
4. Wash hands thoroughly with soap and water after removing gloves.
5. Dispose off gloves and blood-contaminated materials in a biohazard receptacle.

Additional precautions that may not apply to this laboratory exercise include:

1. Wear a lab coat when soiling with side shields if splashing on the face is possible.
2. Wear a mask, goggles or glasses with side shields if splashing on the face is possible.

# Microbial Cell Counting

Microbiological analysis of food, water, milk and air requires quantitative determination i.e. total population of micro-organisms in these substrates. The density of cells, spores/conidia of microorganisms can be measured in the laboratory by several methods either by direct or indirect counts. In the direct microscopic count, a known volume of liquid is added to the slide and the number of microorganisms are counted by examining the slide with the bright-field microscope. For direct microscopic counts Neubauer or Petroff-Hausser counting chamber, breed smears or electronic cell counter (as coulter counter) are used. Various indirect methods are also available for measuring the population density. The two most common indirect methods are plate count and turbidity measurements.

# 1

# MEASUREMENT OF CELLS OF BACTERIA OR FUNGAL CONIDIA BY A COUNTING CHAMBER

## OBJECTIVE

To count and calculate the number of spores or cells in a sample.

### INTRODUCTION

Haemocytometer is a special microscope slide with a counting chamber 0.1 cubic mm originally devised for counting blood cells. It is used for counting the bacteria or fungal spores in a liquid suspension. The counting chamber has a total of nine squares, each of 1 mm × 1 mm engraved over it (Fig. 1.1) but only one  square per field (Fig. 1.1 b) is visible under 100X microscope magnification (10X ocular and 10X objective). A one mm square is divided into 25 medium sized squares (groups) (0.2 mm × 0.2 mm each), each of which is further subdivided into 16 small squares (0.05 × 0.05 mm each), thus a total of 400 squares in 1 mm (Fig. 1.1 c). Each medium-sized square (group) is separated by triple lines, the middle one acts as the boundary. Each large square has a volume of $1 \times 1 \times 0.1$ mm (or $1/10 \times 1/10 \times 1/100$ cm) $= 1/10000$ cm$^3$ or $10^{-4}$ cm$^3$. The cell suspension is introduced into the counting chamber and the total cell number is determined mathematically by counting the number of cells in the chamber.

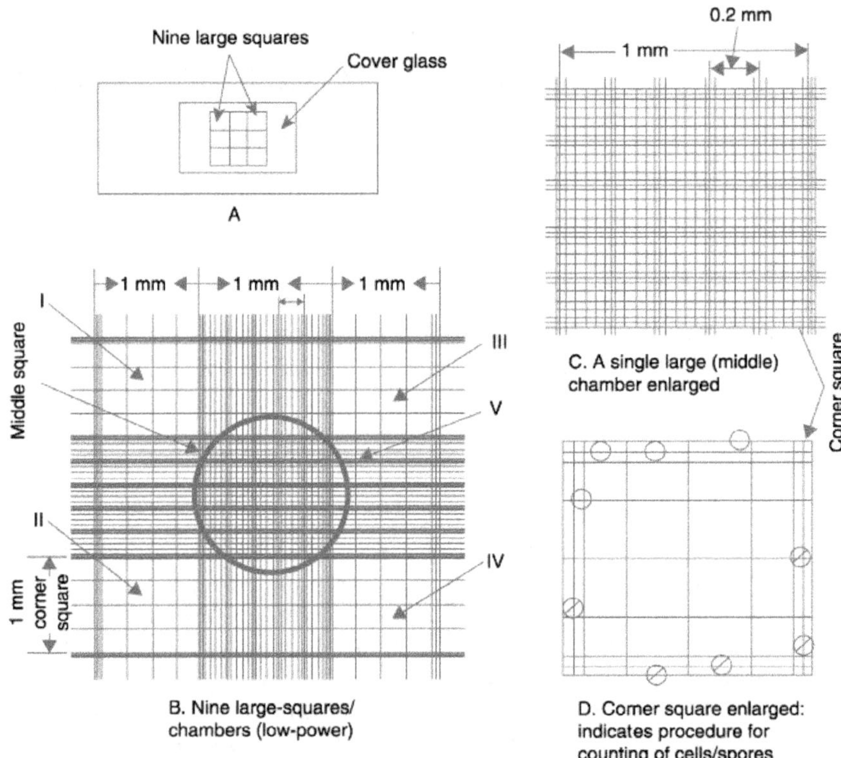



**Figure 1.1** Haemocytometer used to count spores of microorganisms and total count of cells (a) Haemocytometer. (b) Standard haemocytometer chamber. The circle indicates the approximate area covered at 100X microscope magnification (10X ocular and 10X objective). (c) A single large (middle) square enlarged. (d) Corner square enlarged. Count cells on top and left touching middle line (O) and do not count cells touching middle line at bottom and right.

## MATERIALS

### Equipments

Counting chamber (haemocytometer)

### Cultures

Cell or spore/conidial suspension

### Miscellaneous

Pipette

## PROCEDURE

1. Place the cover glass (special cover glass of counting chamber) over the grid carefully and introduce a drop of bacterial or conidial suspension made from liquid culture in-between the coverslip and the grid .
2. Slide the cover glass backwards and forwards until coloured rings are visible as the two surfaces of cover glass and slide come into close contact.
3. Count the bacteria or spores in 4 corner large squares (I, II, III, IV) and in the middle one (V) to have a total count of 200–250.

<table>
<tr><td rowspan="3" style="writing-mode: vertical-rl;">CAUTION</td><td>

- Always count cells/spores on top and left touching middleline of the perimeter of each square (Fig. 1.1d) and do not count spores touching the middle line at bottom and right sides.

- Do not overfill or underfill the chambers or haemocytometer.

- A minimum of 200–250 spores (i.e. 20–25 spores/ square) should be counted to avoid error.

</td></tr>
</table>

## RESULTS

Calculate the number of spores/cells per ml of the suspension mathematically as follows:

## CALCULATION

Number of cells or spores in the five squares counted = y cells

Volume of the grid = 1 mm × 1 mm × 0.1 mm
$$= 0.1 \times 0.1 \times 0.01 \text{ cm}^3 = 10^{-4} \text{ ml}$$

Number of cells in 25 squares = y × 5 cells/$10^{-4}$ ml
$$= y \times 5 \times 10^4 \text{ cells/ml}.$$

REVIEW QUESTIONS

1.  Define (i) Cells (ii) Spores

2.  List the various methods available to count the number of microorganisms in a sample.

# ENUMERATION OF BACTERIA BY SERIAL DILUTION AGAR PLATE TECHNIQUE

## OBJECTIVES

1. To become familiar with serial dilution and plating technique.
2. To be able to differentiate the colony characters of different bacteria.

## INTRODUCTION

This method is based on the principle that when material containing bacteria is cultured, every viable bacterium and develops into a visible colony on a nutrient agar medium. The number of colonies, therefore, are the same as the number of organisms contained in the sample. In this procedure a small measured volume (or weight) is mixed with a large volume of sterile water or saline called the diluent or dilution blank. Dilutions are usually made in multiples of ten and plated on nutrient agar and the number of colonies formed were counted. Using this, number of microorganisms present in the original sample was calculated.

## MATERIALS

### Culture

Sample or bacterial suspension

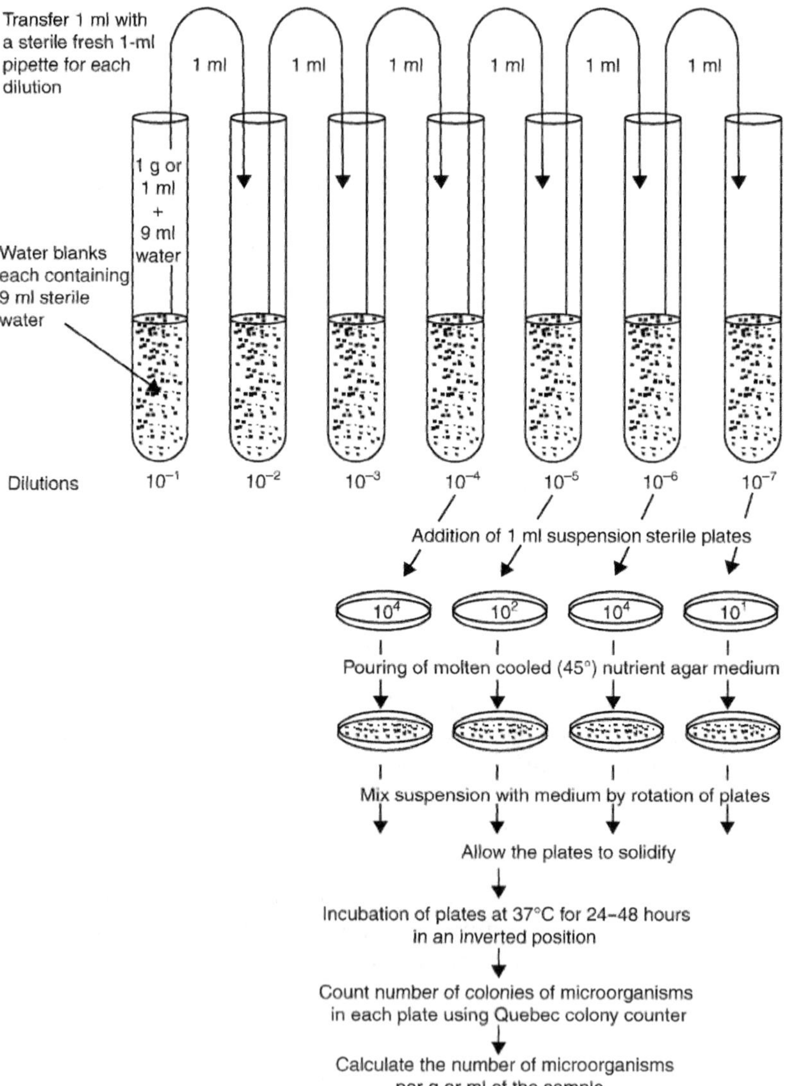


**Figure 2.1**   The standard plate count technique to determine the total number of microorganisms in a sample.

### Media and Solutions

Nutrient agar
Diluent (0.85% NaCl)

### Miscellaneous

Test tubes
Sterile petri plates (12)
Sterile 1-ml pipettes (7)
Colony counter

## PROCEDURE

1. Label the dilution blanks as $10^{-1}$, $10^{-2}$, $10^{-3}$, $10^{-4}$, $10^{-5}$, $10^{-6}$ and $10^{-7}$.
2. Prepare the initial dilution by adding 1 ml or 1 g of the sample into a 9 ml dilution blank (if a solid substance is taken 10 ml of dilution blank should be taken in the first tube) labelled $10^{-1}$ thus diluting the original sample 10 times $(1/(1+9) = 1/10$ and is written $1 : 10$ or $10^{-1})$.
3. Mix the contents by rolling the tube back and forth between your hands to obtain uniform distribution of organisms (cells).
4. From the first dilution, transfer 1 ml of the suspension while in motion, to the dilution blank $10^{-2}$ with a sterile and fresh 1-ml pipette diluting the original specimen/suspension to 100 times $(1/10 \times 1/10 = 1/100$ or $10^{-2})$.
5. From the $10^{-2}$ suspension, transfer 1 ml of suspension to $10^{-3}$ dilution blank with a fresh sterile pipette, thus diluting the original sample to 1000 times $(1 : 1000$ or $10^{-3})$.
6. Repeat this procedure till the original sample has been diluted 10,000,000 $(10^{-7})$ times (Fig. 2.1); every time a fresh sterile pipette should be used.
7. From the appropriate dilutions $(10^{-1}$ to $10^{-7})$ transfer 1 ml or 0.1 ml of suspension while in motion, with the respective pipettes, to sterile petri dishes. Three petri dishes are to be used for each dilution. 1 ml of the diluted sample is transferred into the petri dish.
8. Add approximately 15 ml of the nutrient medium, melted and cooled to 45°C, to each petri plate containing the diluted sample. Mix the contents of each plate by rotating gently to distribute the cells throughout the medium.

## OBSERVATIONS

1. Observe all the plates for the appearance of bacterial colonies.
2. Count the number of colonies in the plates that have colonies in the 30–300 range, by placing each plate one by one on the platform of a Quebec colony counter.

## RESULTS

Calculate the number of bacteria per ml of the original suspension/sample as follows:

$$\frac{\text{Organisms per millilitre}}{\text{Gram of the sample}} = \frac{\text{Number of colonies (average of 3 replicates)}}{\text{Amount plated} \times \text{dilution}}$$

For example, if 60 colonies were counted on a $1 : 10^3$ dilution, then

$$\text{Number of cells/ml} = \frac{60\ \text{colonies}}{1\ \text{ml} \times 10^{-3}} = 60000$$

$$= 6.0 \times 10^4\ \text{bacteria/ml or gram of sample.}$$

9.  Allow the plates to solidify.
10. Incubate these plates in an inverted position for 24–48 hours at 37°C.


## REVIEW  QUESTIONS


1.  What is the advantage of saline when used as diluent?

2.  What do you mean by dilution factor?

3.  What is the use of replicates?

# Microscopic Observation of Microorganisms

A considerable amount of information can be gained by careful microscopic examination of microorganisms. Normally two techniques are employed to study microbial cells. The first method employs examination of living cells and the second employs the study of stained cells. Living organisms can be studied to determine the natural size and shape of cells, cellular arrangement and motility. But the observation of cells in their natural or unstained state is sometimes difficult because of their semitransparency. Therefore, stained preparations of killed microorganisms are more frequently employed.

# 3

# WET MOUNT

## OBJECTIVE

After completing this experiment you will be familiar with preparing wet mounts of the microorganisms.

## INTRODUCTION

Direct examination of living microorganisms (viz. bacteria, protozoa, algae) can be made by two methods: wet mount and hanging drop technique. Both the techniques are very useful in determining size, shape and movement.

## MATERIALS

### Culture

12–18-hour-old broth culture of *Bacillus cereus*
*Spirogyra* sp. (culture or in pond water)

### Equipment

Light microscope

### Miscellaneous

Glass slide
Coverslip
Dropping pipettes

## OBSERVATIONS AND RESULTS

Note the size, shape and characteristics of motility of bacteria (i.e. whether true motility or Brownian movement, or a vibratory or dancing movement of bacteria in suspension due to bombardment by the molecules of water) and other microorganisms observed in the preparations.

## PROCEDURE

1. Transfer a small drop each of the *Bacillus* culture and pond water containing the algae and protozoa, using a fresh pipette for each organism, on the centre of clean glass slides.
2. Handle the cover slip by its edges and place it on the drop.
3. Press the cover slip gently with the end of a pencil.


### REVIEW QUESTIONS

1. How will you differentiate between motility and Brownian movement?

# 4

# HANGING DROP TECHNIQUE FOR DEMONSTRATING MOTILITY OF BACTERIA

## OBJECTIVES

1.  To become familiar with different types of motility exhibited by bacteria.
2.  To be able to differentiate between motility and Brownian movement.

## INTRODUCTION

Hanging drop preparation is useful for microscopic examination of living microorganisms, especially bacteria, without staining them and to see their motility due to flagella.

## MATERIALS

### Culture

12-hour-old broth culture of *Proteus vulgaris*

### Equipment

Light microscope

## OBSERVATIONS

Record your findings on the motility of the bacteria.

## RESULTS

True motility will be shown by bacteria that move swiftly across the microscope.

Miscellaneous

Hanging-drop (cavity) slide
Coverslips
Vaseline/petroleum jelly
Match sticks

## PROCEDURE

1. Clean and flame a hanging-drop slide and place it on the table with the depression uppermost.
2. Spread a little vaseline or petroleum jelly around the cavity of the slide.
3. Clean a coverslip and apply petroleum jelly on each of the four corners of the coverslip, using a match stick.
4. Place the coverslip on a clean paper with the petroleum jelly side up.
5. Transfer one loopful of culture in the centre of the coverslip.
6. Place the depression slide on the coverslip, with the cavity facing down so that the depression covers the suspension.
7. Press the slide gently to form a seal between the coverslip and the slide.
8. Lift the preparation and quickly turn the hanging-drop preparation cover slip up so that the culture drop is suspended.
9. Examine the preparation under low-power objective with reduced light.
10. Switch to the high-power objective and examine the preparation again.

**CAUTION**

- If a bacterial culture growing on a solid medium is to be examined, a loopful of culture should be mixed with a drop of 2% CMC, in the centre of the cover slip.

- The depression slide is inverted over the cover slip in such a way that the suspension does not touch the surface of the concavity at any point.

- The slide and coverslip should be sterilised after the examination is finished.

- Petroleum jelly from the depression slide and cover slip should be removed at the end of the experiment, with xylene.

## REVIEW  QUESTIONS

1. Define the terms: (a) Flagella, (b) Monotrichous flagella, (c) Peritrichous flagella, (d) Polytrichous flagella, (e) Lophotrichous flagella.

2. What are the different types of motility?

# 5

# NEGATIVE STAIN

---

## OBJECTIVES

1. To perform a negative staining procedure and to become familiar with different types of stains.
2. To observe how different stains react with cells.

---

### INTRODUCTION

Morphological stains colour either bacterial cell themselves or their backgrounds to allow a clear microscopic view of cells. Such clear views provide information about cell size, shape and arrangement.

Stains that colour bacterial cells themselves carry a positive charge and are called basic stains. Basic stains colour bacterial cells because they are attracted to the negatively charged cell surface. Basic stains include crystal violet, methylene blue, and safranin.

Stains that colour the surrounding bacterial cells carry a negative charge and are called acidic stains. Acidic stains are repelled by the negatively charged bacterial cell surface and, hence, colour only the background. However this still provides a clear microscopic view, because the bacterial cells are seen in outline. Acidic stains include Congo red, nigrosin, and India ink.

A single acidic stain used to colour the background around cells is called a negative stain. There are two advantages of a negative stain: (1) it allows more accurate determination of cell size and shape, since the procedure requires no heating or staining of cells (which can cause cell shrinkage) and (2) it facilitates the microscopic observation of cells that are difficult to stain, such as spirilli and spirochaetes.

In this exercise, you will use a single acidic stain to determine the cell morphology of several bacterial cultures.

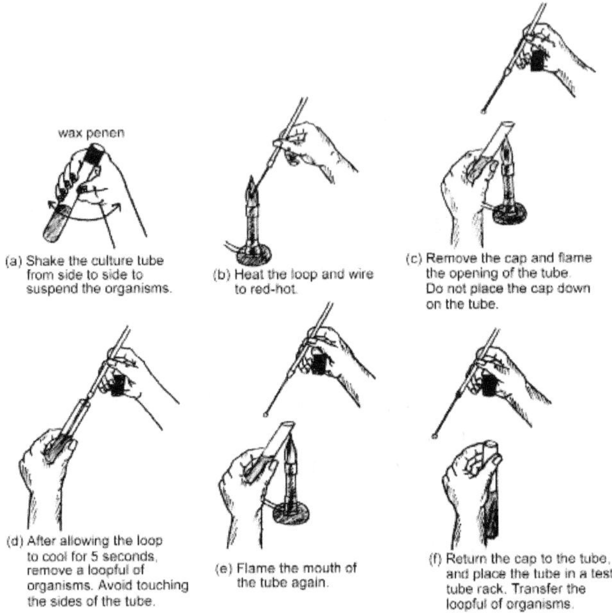



**Figure 5.1**    Aseptic procedure for removing an organism from the broth culture.

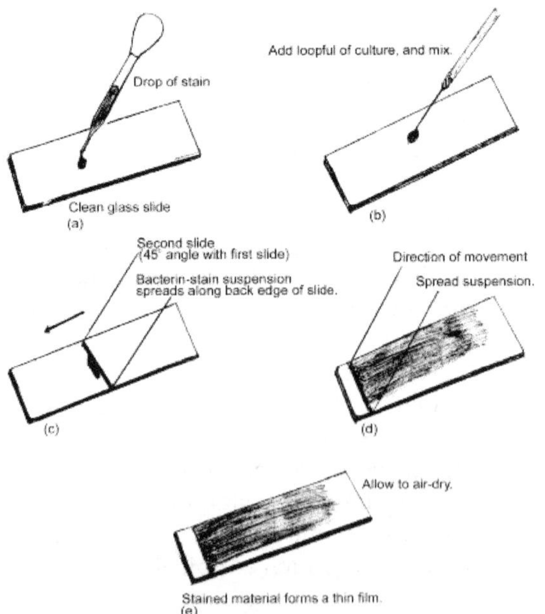



**Figure 5.2**    Negative staining procedure.

## MATERIALS

### Cultures (24–48-hour broth)

*Bacillus cereus* (rod)
*Staphylococcus epidermidis* (coccus)

### Stains

Nigrosin, India ink or Congo red

### Equipment

Light microscope

### Miscellaneous supplies

Bunsen burner
Glass slides
Immersion oil
Inoculating loop
Lens paper

## PROCEDURE

1. Place a drop of nigrosin, India ink, or Congo red near the edge of a clean glass slide.
2. Aseptically obtain a loopful of a broth culture of *Bacillus cereus* by following the steps in Fig. 5.1.
3. Transfer the loopful of culture to the drop of stain on the slide, and mix the culture into the drop, as shown in Fig. 5.2 a, b. Always flame your loop before setting it down.
4. Follow steps c to e in Fig. 5.2 c–e to complete your preparation of a negative stain of *Bacillus cereus*.
5. Repeat steps 1–4 to prepare a negative stain of broth culture of *Staphylococcus epidermidis*.
6. After you have completed both negative stains, examine them using the oil-immersion objective.

## OBSERVATION AND RESULT

Draw the cell shape and arrangements you observed.

1.  Magnification

2.  Cell shape

3.  Cell arrangement

## REVIEW QUESTIONS

1.  Explain why nigrosin, India ink and Congo red do not stain bacterial
    cells?

2.  What are the advantages of negative stain?

# 6

# SIMPLE STAIN

---

## OBJECTIVE

To perform the simple staining procedure to compare morphological shapes and arrangements in bacterial cells.

---

### INTRODUCTION

The use of a single basic stain to colour bacterial cells is called a simple stain. Basic stains employed for this purpose include safranin, crystal violet and methlyene blue. These stains colour the bacterial cells so that they are clearly visible under the microscope. Since this procedure requires the heat-fixation of a smear prior to stain application, it does result in some cell shrinkage.

In this exercise, you will use a single basic stain to colour the cells of several bacterial cultures to reveal their morphological characteristics.

### MATERIALS

Cultures (24–48-hour broth or agar)

*Pseudomonas aeruginosa* (rod)
*Staphylococcus epidermidis* (coccus)

Stains

Crystal violet, methylene blue, or safranin

(a) Flame the loop (or needle) to red-hot to sterilise.

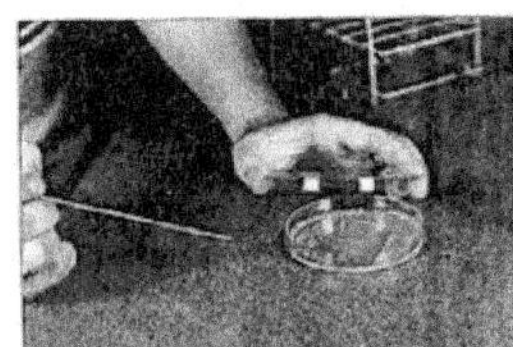

(b) Touch the loop (or needle) to an isolated colony to pick up a pinhead amount of growth.

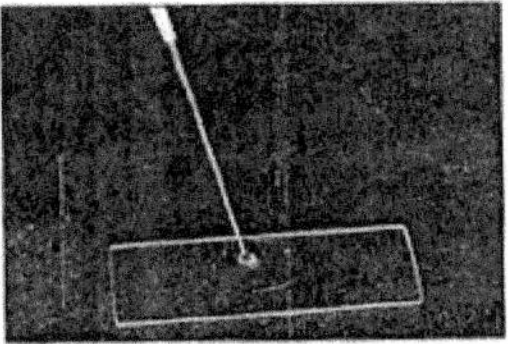

(c) Transfer the growth to a drop of water on a slide and thoroughly mix to obtain a slightly milky colour. This mixture must be air-dried and heat-fixed before staining.

(d) Cover the heat-fixed smear with stain and allow to sit for 60 seconds.

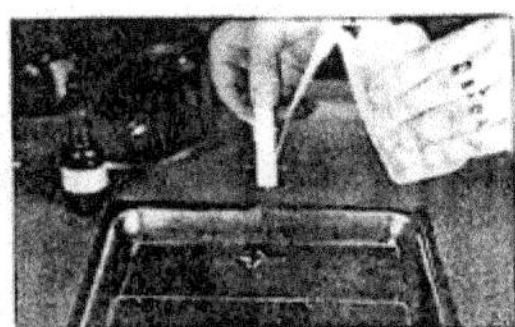

(e) After 60 seconds, wash off the stain with a water rinse. After drying, the stained smear is ready for observation.

**Figure 6.1**   Bacterial smear preparation and simple stain procedure.

### Equipment

Light microscope

### Miscellaneous

Blotting paper
Bunsen burner
Clothespin
Glass slides
Immersion oil
Inoculating loop or needle
Lens paper
Staining tray
Water bottle with tap water

## PROCEDURE

### Smear preparation

1. Aseptically obtain a loopful of broth culture of *Pseudomonas aeruginosa*. If an agar culture is used instead of broth, follow the steps depicted in Fig. 6.1.
2. Transfer the loopful of broth culture to a glass slide, and spread it out into a circle. If an agar culture is used instead of broth, mix the pinhead amount of culture into a drop of water on a glass slide as shown in Fig. 6.1 c. Prepare a mixture that is only slightly milky in colour. Do not prepare a heavy suspension.
3. Allow the slide to air-dry before heat-fixation. Heat gently by passing the slide over the flame several times. After heat-fixation, the slide is ready to stain.
4. Repeat steps 1–3 to prepare a smear of *Staphylococcus epidermidis* and to prepare a smear of a mixture of *Pseudomonas aeruginosa* and *Staphylococcus epidermidis*.

### Simple stain

1. Apply crystal violet, methylene blue, or safranin to the three smears, and let them stand for 60 seconds (Fig. 6.1 d). Cover the entire smear with stain.

## OBSERVATION AND RESULT

Draw your results (shape) from the simple stains.

1.  Magnification

2.  Cell shape

3.  Stain used

4.  Colour of cells

2.  After 60 seconds, gently wash off the stain with tap water (Fig. 6.1 e). Dry the slide with blotting paper, and examine using the oil immersion objective.

## REVIEW  QUESTIONS

1.  What is the purpose of heat-fixation?

2.  What happens if you heat-fix too much?

3.   Why is a thick smear undesirable?

4.   Explain why methylene blue, crystal violet and safranin stain differently form nigrosin, India ink and Congo red?

5.   What are the advantages and disadvantages of simple stain?

# 7

# GRAM STAIN

---

## OBJECTIVES

1. To become familiar with
   a. The chemical and theoretical basis of differential staining procedures.
   b. The chemical basis of the gram stain.
2. To differentiate the two principal groups of bacteria — gram-positive and gram-negative.

---

## INTRODUCTION

Most bacteria possess a cell wall that contains either a thick peptidoglycan layer or a thin peptidoglycan layer with an additional outer membrane composed of lipopolysaccharide (LPS). This chemical difference in bacterial cell wall is identified with the Gram stain. The Gram stain is the stain most frequently used; to identify unknown bacterial cultures, because it yields information on gram reaction, cell size, cell shape, and cell arrangement.

During the Gram-staining procedure, all bacteria are stained purple by crystal violet, the primary stain. Bacterial cells that have a thick peptidoglycan layer retain the crystal violet during subsequent decolourisation and counter stain steps. These bacteria appear purple when viewed under the microscope and are referred to as gram-positive. Bacterial cells that have a thin peptidoglycan layer lose the crystal violet during the decolourisation step, and take up the counter stain safranin. These bacteria appear red when viewed under the microscope and are referred to as gram-negative.

In this exercise, you will use the Gram stain on selected 18–24-hour bacterial cultures.

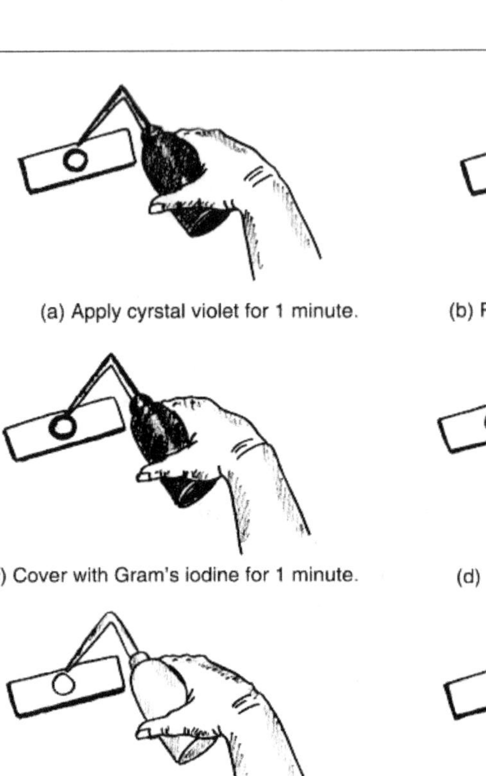

(a) Apply cyrstal violet for 1 minute.

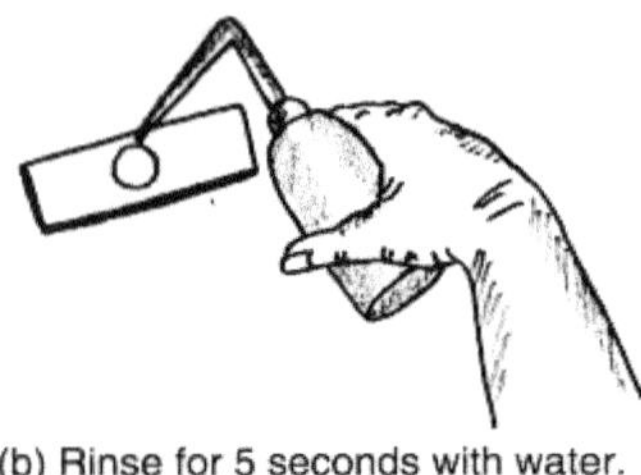

(b) Rinse for 5 seconds with water.

(c) Cover with Gram's iodine for 1 minute.

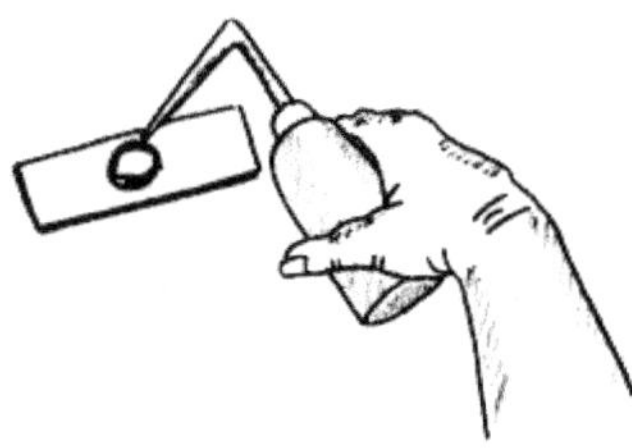

(d) Rinse for 5 seconds with water.

(e) Decolourise with 95% ethanol
for 15–30 seconds.

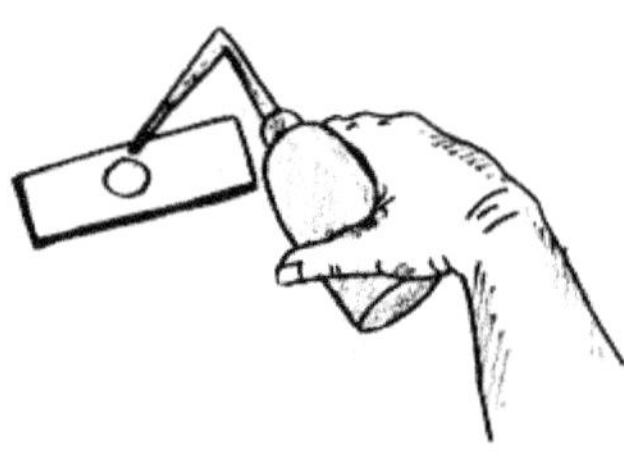

(f) Rinse for 5 seconds with water.

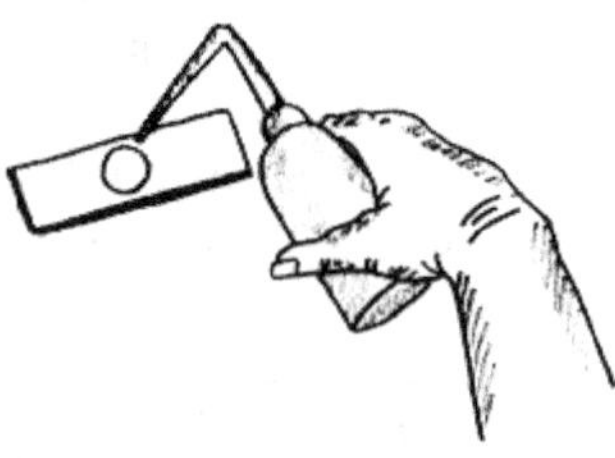

(g) Counterstain with safranin for 1 minute.

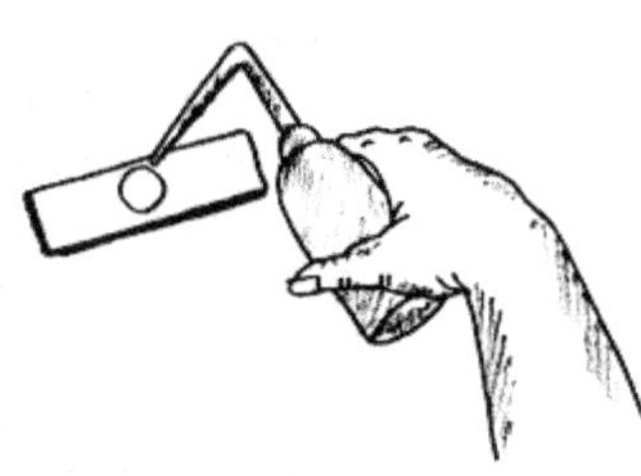

(h) Rinse for 5 seconds with water.

(i) Blot dry with filter paper.

**Figure 7.1**   Gram stain procedure.

## MATERIALS

### Cultures to select from (18–24-hour broth or agar)

*Bacillus cereus* (Gram-positive rod)
*Enterobacter aerogenes* (Gram-negative rod)
*Enterococcus faecalis* (Gram-positive coccus)
*Escherichia coli* (Gram-negative rod)
*Neisseria sicca* (Gram-negative coccus)
*Proteus vulgaris* (Gram-negative rod)
*Pseudomonas aeruginosa* (Gram-negative rod)
*Staphylococcus epidermidis* (Gram-positive coccus)

### Stains

Crystal violet
Gram's iodine
Ethanol (95%)
Safranin

### Equipment

Light microscope

### Miscellaneous

Blotting paper
Bunsen burner
Clothespin
Glass slides
Immersion oil
Inoculating loop or needle
Lens paper
Staining tray
Toothpick
Water bottle with tap water

## PROCEDURE

Select four bacterial cultures from the material list. Although a variety
of bacteria can be used, one from each of the following categories is

## OBSERVATION AND RESULTS

Draw the shape and colour of the cells you have observed under the
microscope.

1.  Magnification


2.  Cell shape


3.  Cell arrangement


4.  Cell colour


5.  Gram reaction

recommended: gram-positive rod, gram-positive coccus, gram-negative rod, gram-negative coccus, and a mixture of gram-positive coccus and gram-negative rod.

1.  Prepare smears of the four selected bacterial cultures and a smear of a mixture.
2.  Using the steps outlined in Fig. 7.1, gram-stain all prepared smears. Follow these steps exactly as outlined. Do not over-decolourise! Tilt the slide and drip alcohol onto the smear until it runs off clear. Stop decolourisation at this point.
3.  After gram-staining, examine all slides using the oil-immersion objective.

**Note**: Avoid viewing areas of the slide where cells are clumped together. Only view areas where individual cells can be seen.

## REVIEW QUESTIONS

1.  Why are contrasting colours important in the Gram stain?

2.  Explain why the alcohol decolourisation step is so critical in Gram stain?

3.  Explain how a Gram stain differs from a: (a)  Negative stain (b)  Simple
    stain.

4.  Why is an 18–24 hour culture necessary for a Gram stain?

# 8

# ACID-FAST STAIN

## OBJECTIVES

1. To become familiar with the chemical basis of the acid-fast stain.
2. To differentiate between acid-fast and non-acid-fast groups of bacteria.

## INTRODUCTION

The acid-fast stain is used to distinguish certain bacteria that contain a high content of the lipid, mycolic acid, in their cell wall. This component makes the cell wall resistant to most stains, but heated carbol fuchsin will penetrate the cell wall, imparting a red colour to cells. It is not removed even when the decolourising agent, acid-alcohol, is added. Bacteria with this characteristic are referred to as acid-fast. The majority of bacteria do not have a high lipid content in their cell wall, so their cells lose the red colour when acid-alcohol is added. They then take up the counterstain methylene blue. These bacteria are referred to as non-acid-fast.

Several pathogenic bacteria can be distinguished by the acid-fast stain, including two species of mycobacteria — *Mycobacterium tuberculosis*, the causative agent of tuberculosis, and *Mycobacterium leprae*, the causative agent of leprosy. An acid-fast stain of sputum is important in the diagnosis of tuberculosis. In addition, certain pathogenic species of the actinomycete genus *Nocardia* are acid-fast, including *Nocardia asteroides*, a causative agent of nocardiosis. The oocysts of the sporozoan parasite *Cryptosporidium* are also acid-fast.

(a) Apply carbolfuchsin to saturate paper, and heat for 5 minutes in an exhaust hood.

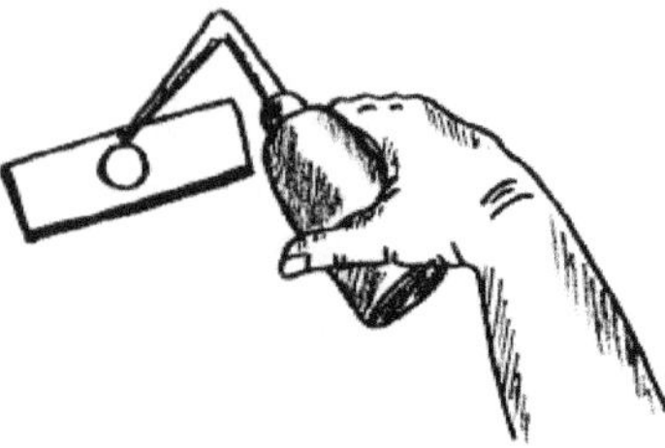

(b) Remove paper, cool, and rinse with water for 30 seconds.

(c) Decolourise with acid-alcohol until pink (10–30 seconds).

(d) Rinse with water for 5 seconds.

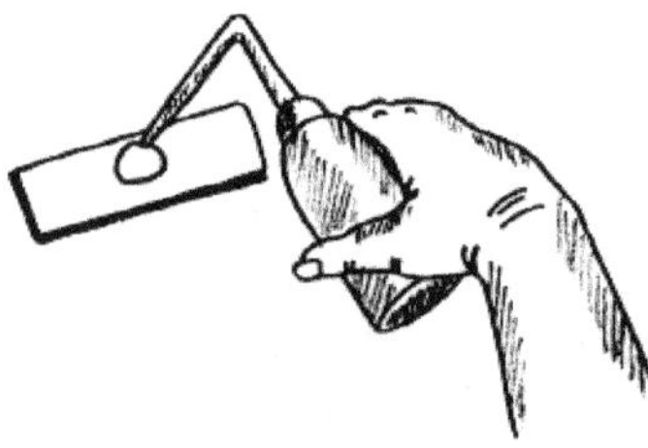

(e) Counterstain with methylene blue for about 2 minutes.

(f) Rinse with water for 30 seconds.

(g) Blot dry with filter paper.

**Figure 8.1**   Acid-fast staining procedure.

## MATERIALS

### Cultures (5–7-day agar)

*Mycobacterium phlei* (acid-fast rod)
*Pseudomonas aeruginosa* (non-acid-fast rod)

### Stains

Carbol fuchsin
Acid-alcohol
Methylene blue

### Equipment

Hot plate (optional, for heating carbol fuchsin)
Light microscope

### Miscellaneous

Blotting paper
Bunsen burner
Clothespin
Egg albumin solution
Glass slides
Immersion oil
Inoculating loop or needle
Lens paper
Staining tray
Water bottle with tap water

## PROCEDURE

1. Mix the culture into a drop of egg albumin solution, instead of water and prepare a smear of *Mycobacterium phlei*, an acid-fast rod, and a smear of *Pseudomonas aeruginosa*, a non-acid-fast rod. This solution will help acid-fast cells to adhere to the glass slide.
2. After smear preparation, follow the steps of the Ziehl–Neelsen acid-fast staining procedure in Fig. 8.1.

OBSERVATION AND RESULT

Draw the results from your acid-fast stains.

1.   Magnification

2.   Cell shape

3.   Cell colour

4.   Acid-fast

**Note**: Carbol fuchsin can be heated using either a hot plate (as depicted) or a Bunsen burner flame. In either case, gently steam only; do not boil. As the stain dries out, add more carbol fuchsin to keep the paper moist. After 5 minutes, remove the paper and continue the steps as outlined.

3.  After staining, examine both slides using the oil-immersion objective.


## REVIEW QUESTIONS

1.  Explain these terms:
    a.   acid-fast.




    b.   non-acid-fast.

2.  Why is heating necessary when we add carbol fuchsin?

3.  Name a few pathogenic acid-fast bacteria?

# 9

# SPORE STAIN

---

## OBJECTIVES

1. To become familiar with the chemical basis of the spore stain.
2. To differentiate between bacterial spores and vegetative cell forms.

---

## INTRODUCTION

Some bacteria produce an internal structure known as an endospore during their life cycle. This structure is produced by the vegetative cell by a process called sporogenesis and is released upon the death of the cell. The resulting free spore is a dormant structure that contains little water and carries out few chemical reactions. Its highly resistant nature is due to two factors: (1) a multilayered outer covering containing peptidoglycan and (2) the presence of a protein-stabilising molecule called dipicolinic acid. These components allow spores to survive adverse conditions that no other living thing could survive. When favourable conditions return, the bacterial spore undergoes germination to yield a vegetative cell.

The detection of endospores is a useful characteristic in the identification of some bacteria, including species of *Bacillus* and *Clostridium*. Several species of these genera are pathogenic — *Bacillus anthracis* causes anthrax; *Clostridium botulinum* causes botulism; *Clostridium tetani* causes tetanus; and *Clostridium perfringens* causes gas gangrene.

When spores are detected in bacteria, their size, shape, and location are useful in identification. Due to their unique physical and chemical characteristics, endospores do not readily stain using ordinary staining procedures. However, basic stains easily colour the vegetative cells that produce endospores. As a result, endospores can be seen in outline against the background of stained vegetative cells. For best

(a) Apply malachite green to saturate paper, and steam for 5 minutes.

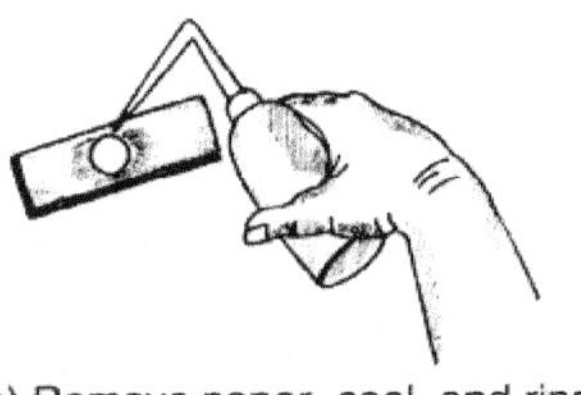

(b) Remove paper, cool, and rinse with water for 30 seconds.

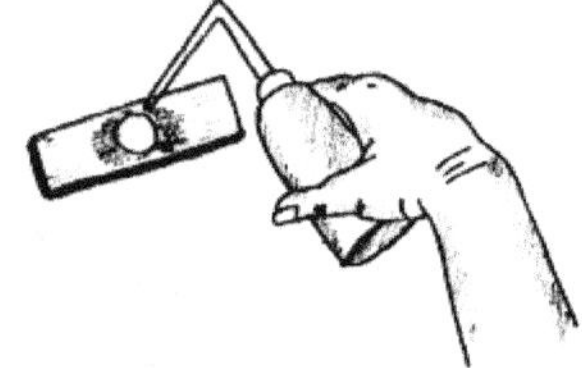

(c) Counterstain with safranin for 60–90 seconds.

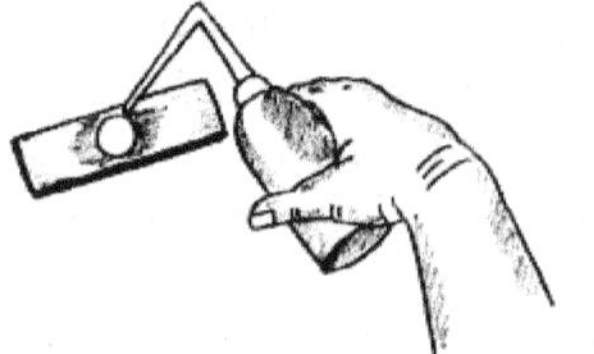

(d) Rinse with water for 30 seconds.

(e) Blot dry with filter paper.

**Figure 9.1**     Schaeffer–Fulton endospore staining procedure.

observation and verification of their presence, spores should be stained using a special procedure called the spore stain.

In a spore stain, the cells are heated in the presence of malachite green. Heating drives the malachite green into the spore, where it is retained even during the water rinse step. When viewed under the microscope, spores are readily visible as green oval-shaped or spherical objects within or outside of vegetative cells. Water removes the malachite green from vegetative cells, allowing them to pick up the counterstain safranin and appear red. Non-spore-forming bacteria appear as red rods with no green, oval-shaped or spherical objects.

In this exercise, you will use the Schaeffer–Fulton endospore stain method to demonstrate the presence of spores in *Bacillus cereus*.


## MATERIALS

### Cultures (4–5 days on nutrient Agar)

*Bacillus cereus* (spore-forming rod)
*Escherichia coli* (non-spore-forming rod)

### Stains

Malachite green
Safranin

### Equipment

Light microscope

### Miscellaneous

Blotting paper
Bunsen burner
Clothespin
Glass slides
Immersion oil
Inoculating loop or needle
Lens paper
Staining tray
Water bath (optional, to heat malachite green)
Water bottle with tap water

## OBSERVATION AND RESULT

1. Magnification

2. Cell shape

3. Vegetative cell colour

4. Spores

5. Colour

6. Location

## PROCEDURE

1. Prepare a smear of *Bacillus cereus*, a spore-forming rod, and a smear of *Escherichia coli*, a non-spore-forming rod.
2. After smear preparation, follow the steps of the Schaeffer–Fulton endospore satin method depicted in Fig. 9.1.

**Note**: Heating the malachite green can be done using a water-bath (as depicted), a hot plate, or a Bunsen burner flame. In either case, gently steam only; do not boil. As the stain dries out, add more malachite green. After 5 minutes, remove the paper and continue the steps as outlined.

3. After staining, examine both slides using the oil-immersion objective.


## REVIEW QUESTIONS

1. Define the terms

   (a) Endospore



   (b) Sporogenesis



   (c) Germination

2.  Explain why sporogenesis is not a form of bacterial reproduction?

3.  How do bacterial endospores differ from mold asexual spores (conidia)?

4.  Why is heat applied to the malachite green in the spore stain? What
    function does water serve in this method?

# 10

## CAPSULE STAIN

---

## OBJECTIVES

1. To become familiar with the chemical basis of capsular stain.
2. To differentiate the capsular material from the bacterial cell.

---

### INTRODUCTION

Some bacteria have cell structures external to the cell wall that are visible under the light microscope after special staining. One of these structures is a capsule, an extracelluar layer surrounding the cell wall that is composed of polysaccharides and polypeptides. Although a capsule is resistant to staining, it can be revealed by using a combination of acidic and basic stains. The acidic stain colours the capsules, while the basic stain colours the cell. The capsule appears as a clear halo around the cell. Non-capsule-forming bacteria do not have a halo around the cell.

Several clinically important bacteria form capsules, including *Klebsiella pneumoniae* and *Streptococcus pneumoniae* both causing bacterial pneumonia. In these and other bacteria, the capsule is considered a virulence factor, since it protects the cell from phagocytosis by white blood cells.

In this exercise, you will prepare a capsule stain of an encapsulated and nonencapsulated culture.

### MATERIALS

Culture (18–24-hour broth)

*Enterobacter aerogenes* (encapsulated rod)
*Alcaligenes denitrificans* (nonencapsulated rod)

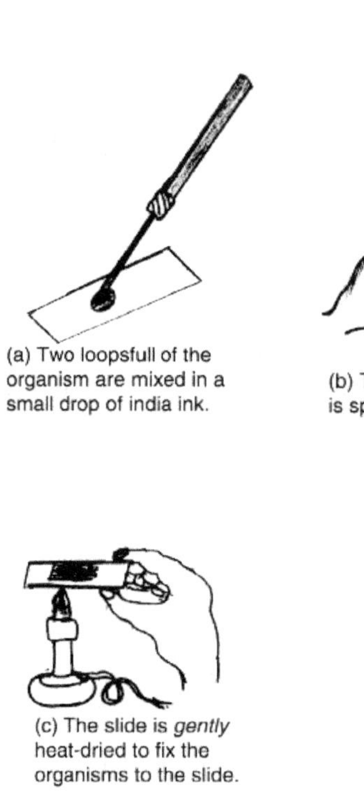


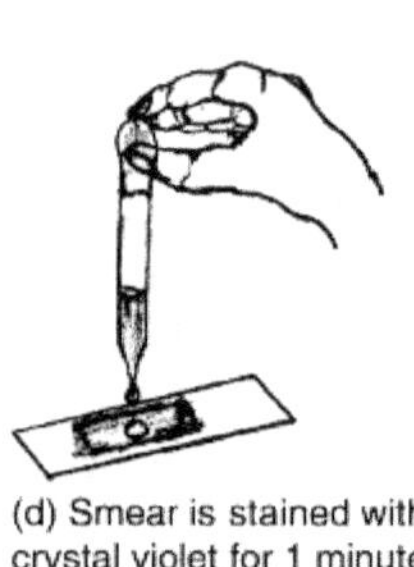


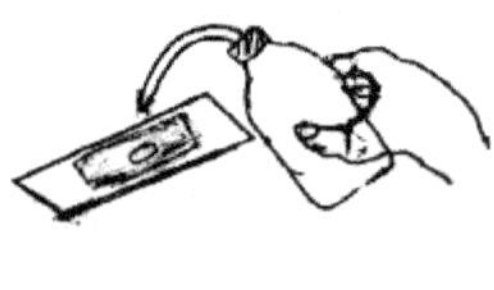


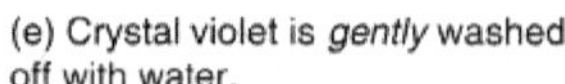

**Figure 10.1**   Procedure for demonstration of capsule.

### Stains

Acidic: India ink
Basic: crystal violet

### Equipment

Light microscope

### Miscellaneous

Blotting paper
Bunsen burner
Clothespin
Glass slides
Immersion oil
Inoculating loop
Lens paper
Staining tray
Water bottle with tap water

## PROCEDURE

### Capsule stain

1. Study the steps for preparing a capsule stain shown in Fig. 10.1. Then carefully follow this procedure as you prepare stain of two cultures: *Enterobacter aerogenes*, an encapsulated rod, and *Alcaligenes denitrificans*, a nonencapsulated rod. Gently heat and rinse with water.
2. When staining is completed, examine both slides using the oil-immersion objective.

## OBSERVATION AND RESULT

Draw the results of your capsule stain

1.   Magnification

2.   Colour of the background

3.   Cell shape

4.   Capsule

5.   Colour

## REVIEW  QUESTIONS

1.  What are bacterial capsules? How do capsules play a role in the establishment of disease?

2.  Why must a combination of basic and acidic stains be used to read a capsule?

3. Name several capsule-forming bacteria and the diseases they cause?

# 11

# SLIDE CULTURE TECHNIQUE

## OBJECTIVES

1. To become familiar with the culturing and staining procedures.
2. To identify fungi.

## INTRODUCTION

The slide culture technique is used to observe morphological characteristics of molds without disturbing the arrangement of spores and conidiogenous cells. The other important application of this technique lies in studying the conidial ontogeny over a period of time in a given area of the preparation.

## MATERIALS

### Culture

Stock culture of a fungus

### Media

Czapek–Dox agar poured and cooled in a sterile petri dish

### Miscellaneous

Sterile petri dish
Glass microscope slide
22 × 30 mm coverslip

## OBSERVATIONS

Examine daily for the appearance of growth of the fungus, and remoisten the filter paper with sterile water when it becomes dry.

Take out the slide from the petri dish at intervals, up to one month and examine microscopically under low-power and high-power objectives for sporulation.

7 cm long sterile applicator sticks
Whatman (qualitative) filter paper
Forceps
Dissecting needle
Transfer needle
Fresh blade
95% alcohol
Sterile distilled water

## PROCEDURE

### Observation of fungal growth

1. Place one or two pieces of filter paper in the bottom of petri dish and moisten with water.
2. Sterilise the above petri dish.
3. Aseptically place two applicator sticks across the filter paper using flamed forceps.
4. Take a clean slide, sterilise it by flaming.
5. Place the sterile slide across applicator sticks.
6. Cut a 10-mm square agar block with a flamed blade.
7. Gently keep the agar block on the centre of the slide in a sterile petri dish.
8. Inoculate a very small portion of the fungus onto each of the four corners of the agar block.
9. Flame a coverslip and place it gently on the surface of the agar block.
10. Replace the lid of the petri dish.
11. Incubate, the side up, at 25°C (or room temperature).

### Preparation of semipermanent mounts

*(A) From coverslip*

1. Lift the coverslip gently from the surface of the agar with forceps.
2. Put a drop of 95% alcohol on the fungus side of the coverslip.
3. Drain off the alcohol.
4. Place a drop of lactophenol cotton blue on a clean slide and put the coverslip over it with the fungus side facing down.
5. Examine the slide microscopically.

## OBSERVATIONS

Examine the stained slide microscopically under low-power and high-power objectives and observe the arrangement of conidia/spores on the conidiogenous hyphae/sporangiophores and identify the fungus.

*(B) From slide*

1. Gently remove an agar block from the slide using a clean blade.
2. Place the agar block in the disinfectant solution.
3. Put a drop of 95% alcohol on fungal growth on the slide.
4. Drain off the alcohol into discard container.
5. Place a drop of lactophenol cotton blue on fungal growth.
6. Place a clean coverslip.

## REVIEW QUESTIONS

1. What is the advantage of slide culture technique?

2. Define

   (i) hyphae

   (ii) spores

3.   What type of stain is lactophenol cotton blue?

# 12

# PREPARATION AND INOCULATION OF GROWTH MEDIA

## OBJECTIVES

To become familiar with

1. The basic nutritional requirements of an organism.
2. The principles associated with routine and special-purpose media for microbial cultivation.
3. The inoculation techniques.

## INTRODUCTION

*Media Preparation*   The cultivation of bacteria (i.e. their growth on a nutrient medium) is necessary for subsequent isolation and identification. A complex medium is one that contains an array of organic nutrients that can grow a variety of bacteria. Media of this type include tryptic soy broth and tryptic soy agar. In this exercise, you will cultivate bacteria using different forms of these media, including broth tubes, agar slants, agar deeps, and agar plates.

*Media Sterilisation*   Before use, the media must be sterilised in an autoclave. Sterilisation is a process that destroys all microbes in a medium. If autoclaving is not done, these microbes will contaminate the culture you introduce into the medium.

*Media Inoculation*   After media preparation and sterilisation, a culture is inoculated (introduced) into each medium. Media inoculation can be done using a variety of sterile instruments such as a loop, needle, swab, or pipette. In all cases, care must be taken to avoid introducing environmental bacteria and fungi into the medium with the culture.

## MEDIA INOCULATION AND INCUBATION

After the different forms of media (broth, slants, deeps, and plates) have been prepared, sterilised and cooled, they are ready for inoculation. Begin by inoculating the two tubes of tryptic soy broth with a culture of *Escherichia coli*. Use an inoculating loop, and follow the procedure given below.

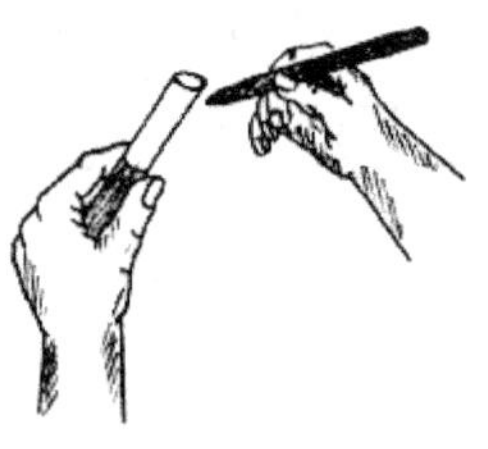

(a) Label the tube to be inoculated with the microorganism used, the date, and your name or initials.

(b) Take the broth culture in one hand.

(c) Take the inoculating loop with your other hand, and flame the entire wire portion to redness.

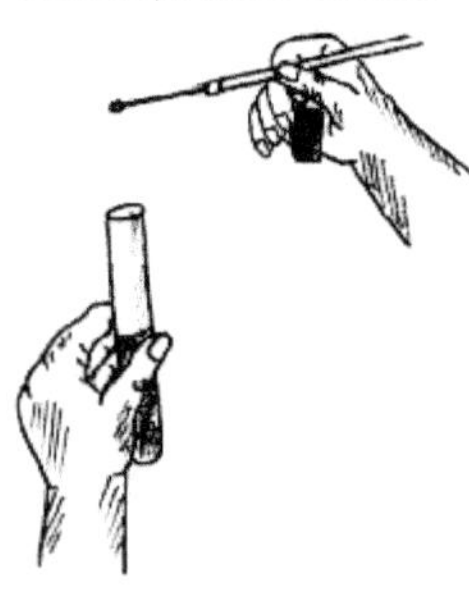

(d) Remove the plug or cap from the tube by grasping it between the fingers of the hand holding the inoculating loop.

(e) Flame the mouth (tip) of the broth culture.

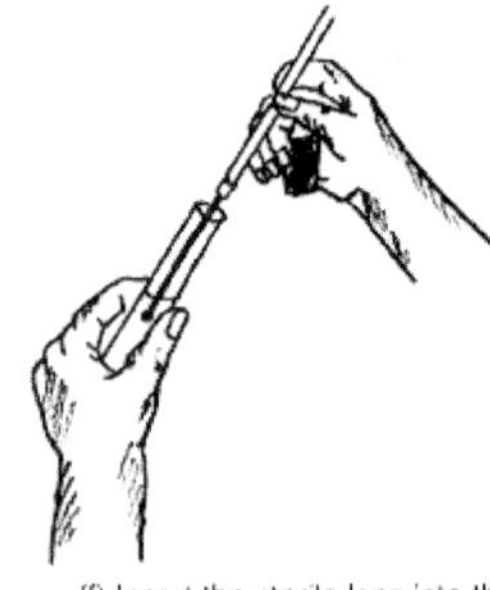

(f) Insert the sterile loop into the broth culture, and obtain a loopful of culture. Withdraw the loop, flame the mouth, and replace the plug or cap. Set the broth culture in a test tube rack. Pick up the tube to be inoculated, remove the plug or cap, and flame the mouth.

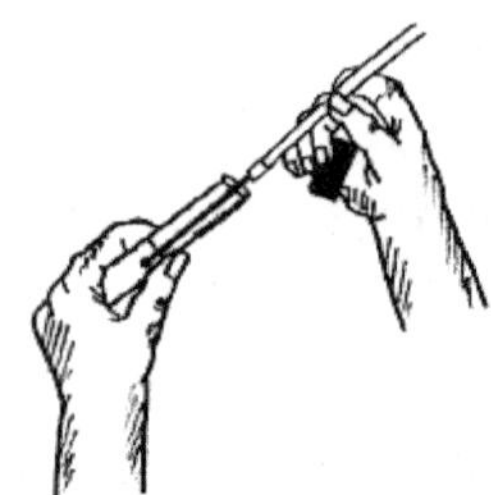

(g) Introduce the loopful of culture by immersing the loop into the sterile broth. Stir lightly.

(h) Withdraw the loop, flame the tip and replace the plug or cap. Set the inoculated tube in the test tube rack.

(i) Flame the inoculating loop again and put it down. Incubate the inoculated tube as directed.

**Figure 12.1**    Broth inoculation.

To prevent contamination of your media, carefully follow the aseptic techniques.

> Prior to culture transfer:
>
> 1. Close doors, this reduces air currents that suspend particles.
> 2. Wash hands, this removes bacteria from your hands.
> 3. Disinfect countertops, this kills bacteria and mold spores in your work area.

> During culture transfer:
>
> 1. Sterilise loop or needle. Hold the loop or needle in a Bunsen burner flame until red-hot.
> 2. Hold test tubes at an upward angle, and flame the opening. This reduces the likelihood of suspended particles that would enter.
> 3. Keep lids on plates when not in use. This reduces the chances of contamination.
> 4. Keep movements in your work area to a minimum. Needless movements create air currents that suspend particles.

*Media Incubation and Examination*   Inoculated media must be incubated to allow time for bacterial growth. After incubation, the microbial growth in tubes and plates is visible and can be examined. During examination, look for contamination by unwanted bacteria and fungi. The presence of more than an occasional contaminant suggests the need for better aseptic technique.

## MATERIALS

### Culture (24-hour broth)

*Escherichia coli*

## MEDIA INOCULATION AND INCUBATION

After the different forms of media (broth, slants, deeps, and plates) have been prepared, sterilised and cooled, they are ready for inoculation. Begin by inoculating the two tubes of tryptic soy broth with a culture of *Escherichia coli*. Use an inoculating loop, and follow the procedure given below.

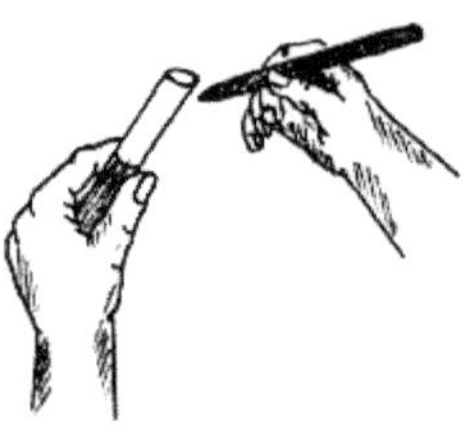

(a) Label the agar slant to be inoculated with the microorganism to be used, the date, and your name or initials.

(b) Take the broth culture in one hand.

(c) Take the inoculating loop with your other hand, and flame the entire wire portion to redness.

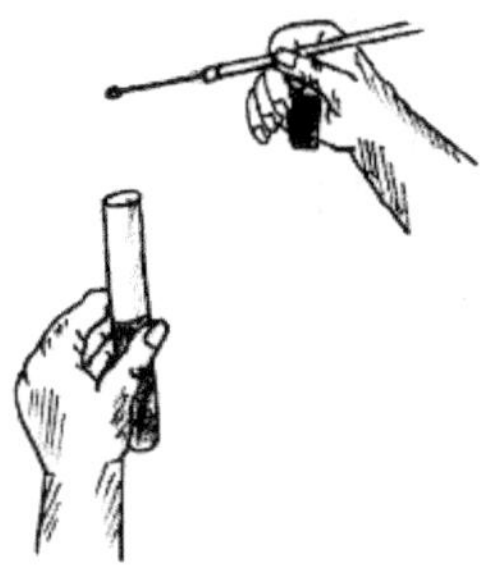

(d) Remove the plug or cap from the tube by grasping it between the fingers of the hand holding the inoculating loop.

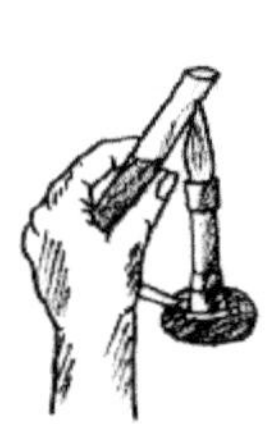

(e) Flame the mouth of the broth culture.

(f) Obtain a loopful of the broth culture. Withdraw the loop, flame the mouth, and replace the plug or cap. Set the tube down, and pick up the agar slant to be inoculated. Remove the plug or cap, and flame the mouth.

(g) Place the loop on the agar slant's surface as its bottom. Move the loop from side to side as you pull it upward out of the tube.

(h) Withdraw the loop, flame the tip and replace the plug or cap. Set the inoculated tube in the test tube rack.

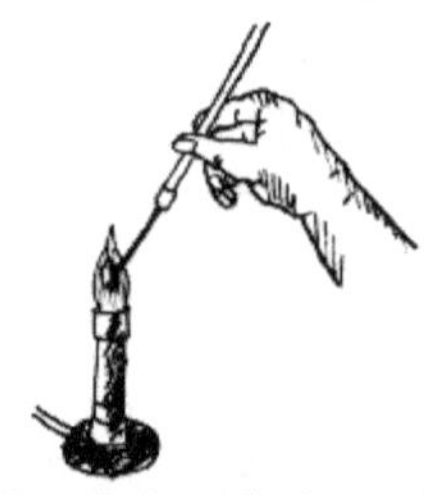

(i) Flame the inoculating loop again and put it down. Incubate the inoculated tube as directed.

**Figure 12.2**   Inoculation on agar slant.

### Media

Tryptic soy agar
Tryptic soy broth

### Equipment

Autoclave
Balance
Hot plate
Incubator (set at 35°C)

### Miscellaneous

Aluminum foil
Bunsen burner
Distilled water
Erlenmeyer flask, 250 ml
Graduated cylinder, 100 ml
Immersion oil
Inoculating loop and needle
Petri dishes, sterile
10-ml pipette with bulb
Spatula
Stirring bar
Test tubes and caps
Test tube rack
Weigh paper

## PROCEDURE

*Media preparation and sterilisation*

1. Following the directions on the bottle, prepare 10 ml of tryptic soy broth in a small flask or beaker. Heat the mixture until the powder dissolves, and then transfer 5 ml to each of the two tubes with a 10-ml pipette. Cap the tubes loosely, and place them in a test tube rack.
2. Following the directions on the bottle, prepare 115 ml of tryptic soy agar in a 250-ml Erlenmeyer flask. Heat the mixture until the powder dissolves and the medium turns clear.

                    

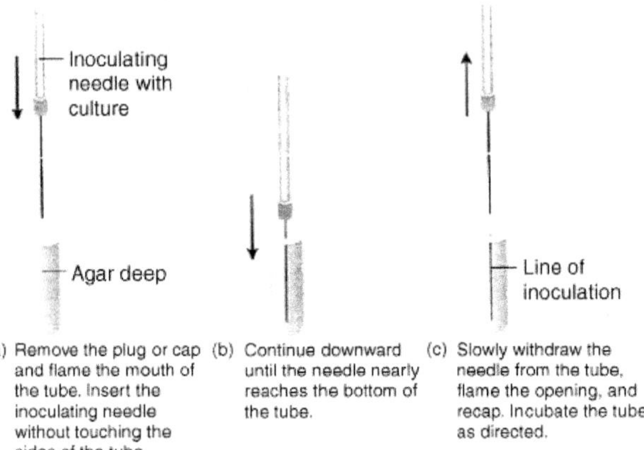


**Figure 12.3**   Stab technique for agar deep cultures.

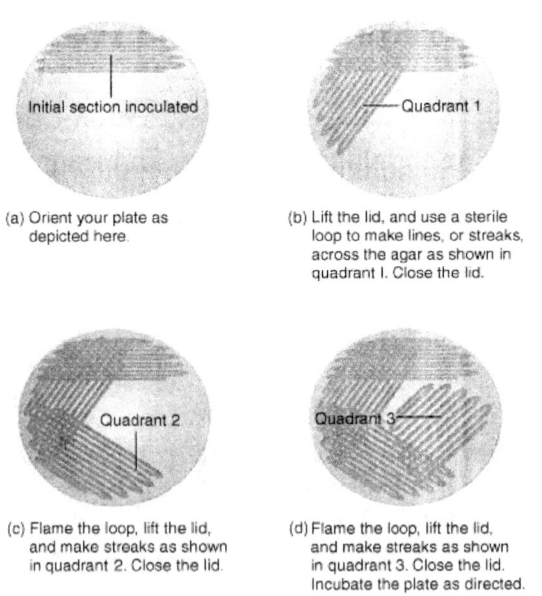


**Figure 12.4**   The streak-plate method.

3.  After the medium has become clear, use a 10-ml pipette to transfer 5 ml to each of the two tubes and 10 ml to each of two other tubes. Cap loosely, and add these four tubes to the two already in the test tube rack. Leave the remaining 85 ml of tryptic soy agar in the flask, and cover with foil. This volume will be used to pour plates after sterilisation.

4.  Label your test tube rack and flask, and place them in the autoclave for sterilisation.

5.  After sterilisation, lean the two tubes with 5 ml of tryptic soy agar against a notebook or similar object. When the medium cools, these tubes will form agar slants. Let the remaining tubes cool upright in the rack. These tubes will form agar deeps.

6.  Let the flask contents cool sufficiently to allow handling without discomfort. When this has occurred, pour the agar into the plates. Do not pour hot agar into a petri dish. This will cause excess condensation on the lid of the petri dish and on the agar surface. Excess moisture may allow the culture to spread across the entire surface of the agar, instead of forming discrete colonies.

7.  When the agar has gelled in tubes and plates, the media are ready to inoculate.

*Media inoculation and incubation*

1.  After the different forms of media (broth, slants, deeps, and plates) have been prepared, sterilised and cooled, they are ready for inoculation. Begin by inoculating the two tryptic soy broth with *Escherichia coli* using an inoculating loop. Follow the procedure outlined in Fig. 12.1.

2.  Inoculate the two tryptic soy agar slants with *Escherichia coli* using an inoculating loop. Follow the procedure outlined in Fig. 12.2.

3.  Inoculate the two tryptic soy agar deeps with *Escherichia coli* using an inoculating needle. The inoculation of deep agar culture is depicted in Fig. 12.3.

4.  Inoculate the three plates of tryptic soy agar with *Escherichia coli* using the method depicted in Fig. 12.5. Spread the culture over several quadrants of the plate using the streak-plate method depicted in Fig. 12.4.

(a) Flame the opening of the broth tube.

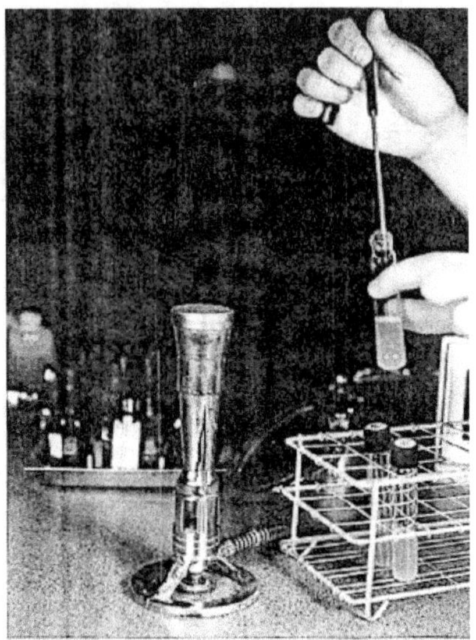

(b) Insert a sterile loop into the broth culture.

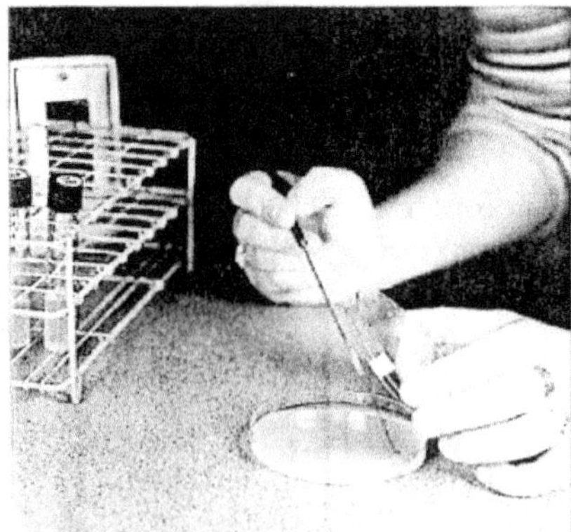

(c) Transfer the culture to a section of an agar plate by
rubbing the loop back and forth across the surface.

**Figure 12.5**   Inoculation on agar plate.

5.  After inoculation, incubate the plates and tubes for 48–72 hours in a laboratory incubator (Fig. 12) set at 35°C.


*Media examination*

1.  After incubation, examine cultures for growth.

**Note**: The assessment of growth in tubes and plates can be aided by comparing inoculated media with uninoculated media. Begin your assessment with broth tubes. Broth tubes should appear cloudy or turbid, when compared to the clear broth in uninoculated tubes.

2.  Growth on slants should be evident extending away from the line of inoculation.
3.  In agar deeps, growth should occur along the needle line of inoculation from top to bottom.
4.  Growth on agar plates should appear as distinct colonies in the second or third quadrant, each colony having the same appearance. Good separation of colonies is essential for two reasons: (a) to confirm the presence of only one species of bacteria (a pure culture) and (b) to determine the specific characteristics of isolated colonies, if colonies are not well-separated, you might consider inoculating another series of plates in an attempt to improve your technique.
5.  After examining all plates and tubes for *Escherichia coli* growth, go back and inspect them for signs of contamination. Look for any type of growth that appears different from that of the culture you inoculated. The absence of contamination indicates a good aseptic technique.


## REVIEW QUESTIONS

1.  Why is it essential that media be sterile prior to use?

## OBSERVATION AND RESULT

Observe the growth and record it

1.   Growth (+ or –)

2.   Contaminants

3.   Description of growth

2.  Why must agar be cooled prior to pouring into plates?

3.  Why are the inoculating loop and needle flamed before and after use?

4.  What is a contaminant? How does it gain entry into your culture? How
    do you protect your lab culture from a contaminant?

5. How does one determine if growth has occurred in broth?6. Why is streak-plating such an essential procedure in the isolation and characterisation of bacteria?

6. Why is streak-plating such an essential procedure in the isolation and characterisation of bacteria?

# 13

# CULTURE CHARACTERISATION OF BACTERIA

## OBJECTIVE

To study the nutritional requirements of bacteria.

## INTRODUCTION

When a single bacterial culture is grown using different forms of media (broth, slants, deeps, and plates), it displays a collective pattern of growth that is unique to its species. This unique pattern of growth is referred to as its culture characteristics. An organism's culture characteristics can help to distinguish it from other organisms, since each bacterial species typically has a unique pattern of growth. Although useful, culture characteristics alone cannot be relied upon to identify the many species of bacteria. They must be combined with staining reactions and biochemical characteristics.

In this exercise, you will use different forms of media to determine the culture characteristics of five known bacteria. You will also be given one of these five bacteria as an unknown to identify.

## MATERIALS

### Cultures (24–48-hour broth)

*Bacillus cereus*
*Micrococcus luteus*
*Proteus vulgaris*
*Pseudomonas aeruginosa*
*Staphylococcus epidermidis*

## OBSERVATION AND RESULT

Fill the following table from your observations.

| Organism | Colony morphology | Growth on slants | Growth in deeps | Growth in broth | Motility test agar |
|---|---|---|---|---|---|
|  |  |  |  |  |  |
|  |  |  |  |  |  |
|  |  |  |  |  |  |
|  |  |  |  |  |  |

## Media

6 tubes tryptic soy broth
6 slants tryptic soy agar
6 deeps tryptic soy agar
6 plates tryptic soy agar
6 tubes motility test agar

## Equipment

Incubator (set at 35°C)

## Miscellaneous

Bunsen burner
Inoculating loop and needle
Test tube rack

## PROCEDURE

*Media inoculation and incubation*

1. *Inoculate 6 tryptic soy broth tubes*  5 tubes with the known cultures (*Bacillus cereus, Micrococcus luteus, Proteus vulgaris, Pseudomonas aeruginosa,* and *Staphylococcus epidermidis*) and one tube with the unknown culture (one of the previous five cultures, but designated by number only).
2. *Inoculate 6 tryptic soy agar slants*  5 slants with the known cultures and 1 slant with the unknown culture.
3. *Inoculate 6 tryptic soy agar deeps*  5 deeps with the known cultures and 1 deep with the unknown culture. Use an inoculating needle and a straight line of inoculation almost to the bottom.
4. *Inoculate 6 motility test agar tubes*  5 tubes with the known cultures and 1 tube with the unknown culture. Use an inoculating needle and a straight line of inoculation two-thirds of the way down. Motility test agar is used to determine whether a culture is motile or not.
5. *Inoculate 6 tryptic soy agar plates*  5 plates with the known cultures and 1 plate with the unknown culture. Use an inoculating loop and the streak-plate method.

6. Incubate all inoculated tubes and plates in a 35°C incubator for 48–72 hours.

*Media examination*

1. After incubation, examine all plates and tubes for growth.

**Note**: To aid in the interpretation of growth, use an uninoculated plate or tube for comparison. Begin your examination with the six broth tubes. Among the growth patterns in broth determine which pattern is displayed by each culture and record your determination.

2. Continue examining the growth by inspecting slants, deeps, motility test agar, and plates. Again, consult the growth patterns in the figure to determine which pattern is displayed in the appropriate medium by each culture and record your results.
3. After inspecting your cultures and completing the laboratory report table, identify your unknown culture.

## REVIEW QUESTIONS

1. Define these terms

    a. Colony

    b. Pigmentation

    c. Facultatively anaerobic

    d. Pellicle

# 14

# BIOCHEMICAL TESTS USED TO IDENTIFY BACTERIA

## OBJECTIVE

To identify an unknown organism.

## INTRODUCTION

Although culture and staining characterisation of bacteria provide a substantial amount of information, these techniques are not sufficient by themselves for the identification of bacteria. The results of staining and culturing must be combined with the results of biochemical tests to definitively identify bacteria. Biochemical tests evaluate the metabolic properties of a bacterial isolate. After a number of biochemical tests have been performed, the combination of test results forms a biochemical pattern for an isolate, which is unique for each species.

In this exercise, you will perform eight biochemical tests on known bacterial cultures. You will use two cultures for each test, a culture known to have a positive result and a culture known to have a negative result. This will familiarise you with either test result, allowing you to correctly interpret biochemical test results.

## MATERIALS

Cultures (24–48-hour agar or broth)

*Alcaligenes faecalis*
*Enterobacter aerogenes*

*Enterococcus faecalis*
*Escherichia coli*
*Proteus vulgaris*
*Pseudomonas aeruginosa*
*Staphylococcus epiderimidis*

## Media

1 plate tryptic soy agar
2 slants tryptic soy agar
6 tubes oxidation-fermentation (O-F) glucose medium
2 tubes nitrate broth (with Durham tube)
2 tubes methyl red-Voges Proskauer (MR-VP) medium
4 tubes sulphide indole motility (SIM) medium
2 tubes lactose broth (with Durham tube)

## Equipment

Incubator (set at 35°C)

## Reagents

Hydrogen peroxide (3%)
Kovac's reagent
Methyl red (pH Indicator)
Oxidase reagent

## Miscellaneous

Bunsen burner
Inoculating loop and needle
Mineral oil (sterile)
Pasteur pipette with bulb
Test tube rack

## PROCEDURE

### Inoculation and incubation

1. *Catalase test*   Inoculate 2 tryptic soy agar slants, one with *Enterococcus faecalis*, and the other with *Staphylococcus*

*epidermidis*. Using an inoculating loop, make a back-and-forth streak across the slant surface.

2. *Denitrification test*   Using an inoculating loop, inoculate 2 nitrate broth tubes, one with *Alcaligenes faecalis*, and the other with *Pseudomonas aeruginosa*.

3. *Hydrogen sulphide (H₂S) production*   Inoculate 2 SIM tubes, one with *Escherichia coli*, and the other with *Proteus vulgaris*. Using an inoculating needle, stab the agar with a single in-and-out motion.

4. *Indole production*   Inoculate 2 SIM tubes, one with *Enterobacter aerogenes*, and the other with *Escherichia coli*. Using an inoculating needle, stab the agar with a single in-and-out motion.

5. *Lactose utilisation*   Using an inoculating loop, inoculate 2 lactose broth tubes, one with *Escherichia coli*, and the other with *Proteus vulgaris*.

6. *Methyl red test*   Using an inoculating loop, inoculate 2 MR-VP tubes, one with *Enterobacter aerogenes*, and the other with *Escherichia coli*.

7. *Oxidase test*   With a wax pencil, draw a line down the centre of a tryptic soy agar plate. Using an inoculating loop, inoculate one half of the plate with *Escherichia coli* and the other half with *Pseudomonas aeruginosa*. Use a back-and-forth streak across the surface of the agar.

8. *Oxidation-fermentation (O-F) glucose test*   Using an inoculation needle, inoculate 2 tubes of O-F glucose with *Alcaligenes faecalis*, 2 tubes with *Escherichia coli*, and 2 tubes with *Pseudomonas aeruginosa*. After inoculation, cover one tube in each pair with a 2 cm layer of sterile mineral oil.

**Note**: Mineral oil can be poured directly into the tube without using a pipette.

9. Incubate all tubes and plates at 35°C for 24–48 hours, except MR-VP. MR-VP tubes require a minimum of 72 hours of incubation.

## Reading test results

1. *Catalase test*   Using a Pasteur pipette, place a few drops of 3% hydrogen peroxide onto each slant culture. Watch for immediate signs of bubbling, which represent a positive test; the absence of bubbles indicates a negative test. A slide test can be done by

mixing a small amount of culture into a drop of water on a glass slide. The hydrogen peroxide is then added to the drop.

*Staphylococcus epidermidis* is catalase-positive,
*Enterococcus faecalis* is catalase-negative.

2. *Denitrification test*   Note the small inverted tube at the bottom of the medium. This tube, called a Durham tube, is designed to collect gas. Read this test by looking for gas bubbles in the Durham tube (nothing needs to be added). Nitrate broth contains potassium nitrate. Denitrification by bacteria converts the nitrate to nitrogen gas. Gas bubbles in the Durham tube therefore represent a positive test. The absence of bubbles represents a negative test.

| Positive test | Negative test |
|---|---|
| Nitrate reductase | no nitrate |
| $NO_3 \longrightarrow N_2$ (gas) | $NO_3 \longrightarrow NO_3$ (no gas) |
| (bubbles in Durham tube) | (no bubbles in Durham tube) |
| e.g. *Pseudomonas aeruginosa* | e.g. *Alcaligenes faecalis* |

3. *Hydrogen sulphide ($H_2S$) production*   Examine each SIM tube for the presence of black colour. A black colour indicates the production of $H_2S$ which combines with the peptonised iron in the SIM medium. The resulting FeS causes blackening of the medium, and represents a positive test. The absence of a black colour indicates a negative test.

   *Proteus vulgaris* is positive
   *Escherichia coli* is negative.

4. *Indole production*   Use a dropper to place 5 drops of Kovac's reagent onto the SIM agar in each tube. If the amino acid tryptophan has been broken down by the enzyme tryptophanase to form indole, the Kovac's reagent will combine with the indole to form a red colour. A red colour in the Kovac's reagent at the top of the agar represents a positive test.

   *Escherichia coli* is indole-positive.
   *Enterobacter aerogenes* is indole-negative.

5. *Lactose utilisation*   When examining these tubes look for a colour change in the broth and gas in the Durham tube. Lactose broth contains the sugar lactose and the pH indicator phenol red. When lactose is utilised, acids or acids and gas are produced. The acid causes the pH to decrease, turning the phenol red from red to yellow. The gas collects in the Durham tube. Therefore, a yellow colour with bubbles in the Durham tube represents a positive test. No significant colour change and absence of bubbles in the Durham tube represent a negative test.

   *Escherichia coli* is positive,
   *Proteus vulgaris* is negative.

6. *Methyl red test*   Using a Pasteur pipette, add 10 drops of methyl red pH indicator to each tube. Swirl the tube gently to mix the drops into the broth. Examine each tube for colour change. Bacteria that produce many acids from the breakdown of dextrose (glucose) in the MR-VP medium cause the pH to drop to 4.2. At this pH, methyl red is red. A red colour represents a positive test. Bacteria that produce fewer acids from the breakdown of glucose drop the pH to only 6.0. At 6.0, methyl red is yellow. A yellow colour represents a negative test.

   *Escherichia coli* is methyl red-positive.
   *Enterobacter aerogenes* is methyl red-negative.

7. *Oxidase test*   Drop 1–2 drops of oxidase reagent onto colonies of broth cultures. Watch for a gradual colour change from pink, to light purple, and then to dark purple within 10–30 seconds. Such as colour change indicates the presence of the respiratory enzyme cytochrome *c* oxidase and represents a positive test. No colour change in this period indicates a negative test.

   *Pseudomonas aeruginosa* is oxidase-positive.
   *Escherichia coli* is oxidase-negative.

8. *Oxidation-fermentation (O-F) glucose test*   In these tubes, you will look for colour changes in the medium. O-F glucose medium contains the sugar glucose and the pH indicator bromothymol blue . This pH indicator is green at the initial pH of 6.8, but turns to yellow at a pH of 6.0. If glucose is utilised, acids are produced and the pH drops, causing the bromothymol blue to turn from green to yellow. If both tubes (with and without oil) turn yellow, the test organism is considered a facultative anaerobe, able to use glucose in the presence or absence of

## OBSERVATION AND RESULT

Record your results for the biochemical tests.

| Biochemical test | Reagent added | Observation | Interpretation |
| --- | --- | --- | --- |
| Catalase test | | | |
| Denitrification test | | | |
| $H_2S$ production | | | |
| Indole production | | | |
| Lactose utilisation | | | |
| Methyl red test | | | |
| Oxidase test | | | |
| O-F glucose test | | | |

oxygen. If only the tube without oil turns yellow, the test organism is considered an aerobe, able to use glucose only when oxygen is present. No change in either tube indicates that the test organism is unable to utilise glucose.

*Escherichia coli* is a facultative anaerobe, *Pseudomonas aeruginosa* is an aerobe, and *Alcaligenes faecalis* is nonreactive (inert) on glucose.

# 15

# ASSESSING ANTIBIOTIC EFFECTIVENESS
# THE KIRBY–BAUER METHOD

## OBJECTIVES

1. To test the sensitivity of an organism against different antibiotics.
2. To prescribe a certain antibiotic to treat bacterial diseases.

### INTRODUCTION

Among the various chemical agents used to control microbial growth, antibiotics are unique because they are selective in their action, that is, they specifically target bacterial cells. For this reason, they can be introduced into the human body to treat diseases, with minimal effects on human cells. Since the discovery of penicillin by Alexander Fleming over 60 years ago, antibiotics have become a standard method used by physicians to treat bacterial diseases.

### TYPES OF ANTIBIOTICS

Since the discovery of penicillin, many other useful antibiotics have been developed. Each antibiotic has a specific mechanism of action against bacteria. In some cases, the action is broad-spectrum, or effective against wide varieties of bacteria. In others, the action is narrow-spectrum, or effective against only certain bacteria. Table 15.1 lists 10 selected antibiotics, their effect on cells, and their spectrum of activity.

(a) Dip a cotton-tipped swab into a broth culture.

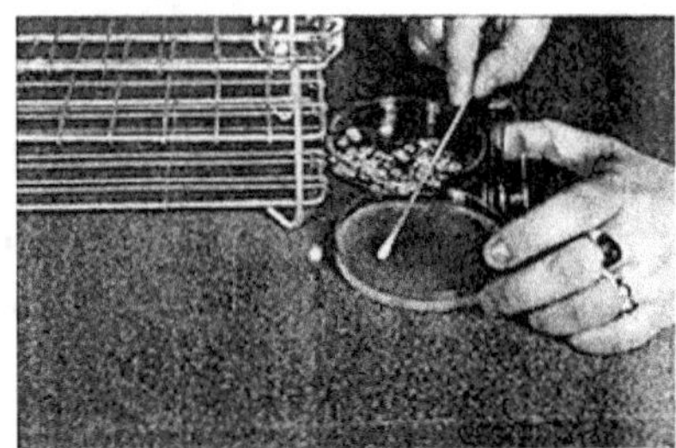

(b) Spread the culture over the entire plate. Dip and spread two more times.

(c) Sterilise forceps by dipping the end in alcohol and then flaming.

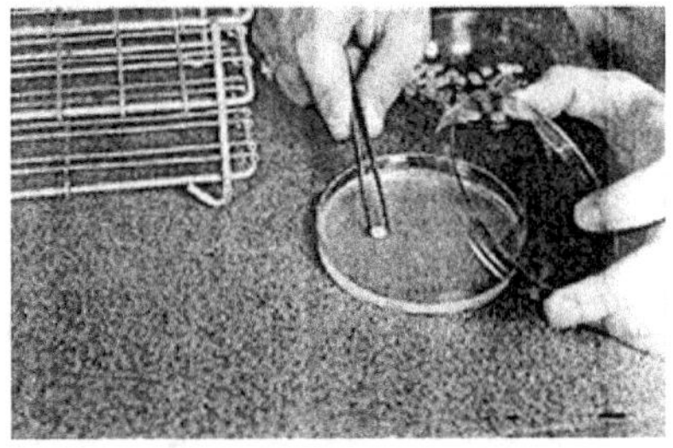

(d) Pick up antibiotic disc, and place on inoculated plate, repeat for four other antibiotics.

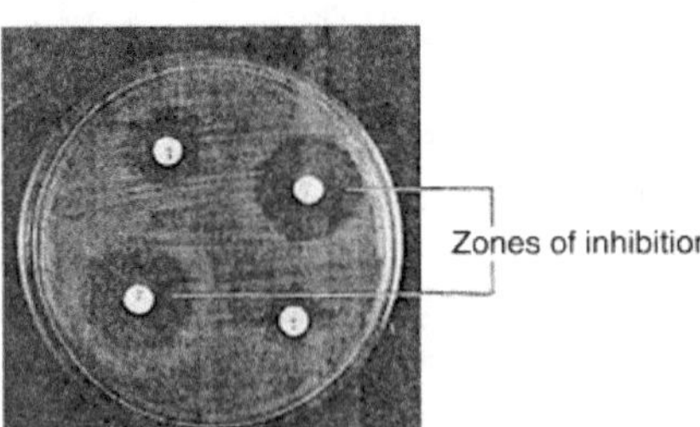


(e) After incubation, examine plates for zones of inhibition, indicative of antibiotic effectiveness.

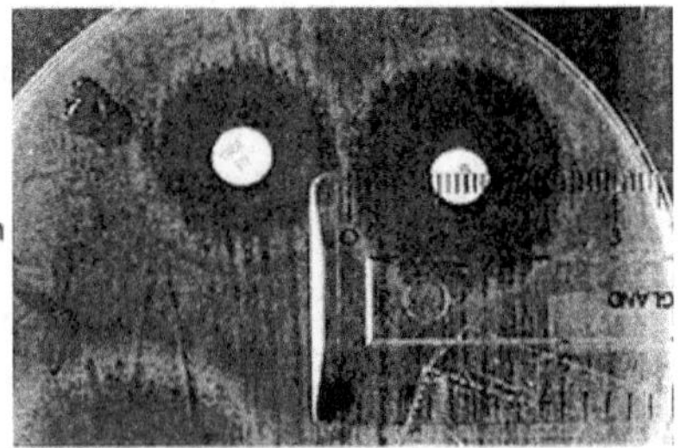

(f) Measure zones of inhibition to the nearest millimetre mm, and compare to the interpretive standard. This zone measures 24mm.

**Figure 15.1**    An outline of Kirby–Bauer antibiotic sensitivity test.

**Table 15.1** Antibiotics used to Treat Bacterial Infections

| Antibiotic | Effect on cells | Spectrum of activity |
| --- | --- | --- |
| Ampicillin | Inhibits cell wall synthesis | Broad-spectrum antibiotic [effective against Gram (+) cocci and some Gram (-) bacterial] |
| Bacitracin | Inhibits cell wall synthesis | Narrow-spectrum antibiotic [effective against Gram + bacterial] |
| Chloramphenicol | Inhibits protein synthesis | Broad-spectrum antibiotic [effective against Gram (+) and Gram (-) bacteria] |
| Erythromycin | Inhibits protein synthesis | Narrow-spectrum antibiotic [effective against Gram (+) bacteria] |
| Gentamicin | Inhibits protein synthesis | Broad-spectrum antibiotic [effective against Gram (+) and Gram (-) bacteria] |
| Penicillin G | Inhibits cell wall synthesis | Narrow-spectrum antibiotic [effective against Gram (+) cocci] |
| Polymymin B | Disrupts cell membrane | Narrow-spectrum antibiotic [effective against Gram (+) rods] |
| Streptomycin | Inhibits cell wall synthesis | Broad-spectrum antibiotic [effective against Gram (+) and Gram (-) bacteria] |
| Tetracycline | Inhibits protein synthesis | Broad-spectrum antibiotic [effective against Gram (+) and Gram (-) bacteria] |
| Vancomycin | Inhibits cell wall synthesis | Narrow-spectrum antibiotic [effective against Gram (+) bacteria] |

## EVALUATING EFFECTIVENESS

When a disease-causing bacterium is isolated from a patient, the physician must determine which antibiotic to administer for treatment. The most widely used method to evaluate the effectiveness of antibiotics against specific bacteria is the Kirby–Bauer method. In the method outlined in Fig. 15.1, Mueller-Hinton agar is inoculated with a culture of a bacterial isolate. After inoculation, antibiotic discs are placed on the agar surface. Plates are incubated to allow for bacterial growth and then inspected for zones of inhibition around antibiotic discs. Zones of inhibition are measured in millimetres and compared to an interpretive standard to determine if the isolate is susceptible or resistant to the antibiotic. Antibiotics that the organism is susceptible to are candidates for use in treating the patient.

## MATERIALS

### Cultures (24-hour in tryptic soy broth)

*Bacillus cereus*, gram-positive rod
*Escherichia coli*, gram-negative rod
*Pseudomonas aeruginosa*, gram-negative rod
*Staphylococcus aureus*, gram-positive coccus

### Media

Mueller-Hinton agar plates, 4 mm thick
(25 ml/plate)

### Chemicals and reagents

Antibiotic discs (loose in sterile petri dish)
Ampicillin
Bacitracin
Chloramphenicol
Erythromycin
Gentamicin
Penicillin G
Polymyxin B
Streptomycin
Tetracycline
Vancomycin
Ethanol, 70%

### Equipment

Incubator (35°C)

### Miscellaneous

Beaker, 250 ml
Bunsen burner
Cotton-tipped swabs, sterile
Disposable gloves
Forceps
Ruler, millimetre
Test tube rack

## PROCEDURE

### Preparation of plates

1.  Dip a sterile cotton-tipped swab into one of the broth cultures, and use it to inoculate a Mueller–Hinton agar plate using the procedure depicted in Fig. 15.1. Inoculation of the plate in this way ensures a lawn of bacterial growth after incubation. Repeat this inoculation procedure for a second plate using the same organism. Label both plates.
2.  Repeat step 1 for the other three cultures. You should now have a total of eight inoculated and labelled plates, two for each culture. After inoculation, allow all plates to dry for 15 minutes before proceeding to the next step.
3.  Pour some 70% ethanol into a 250 ml beaker.
    a.  Dip your forceps into the alcohol, and then pass the forceps over the Bunsen burner flame to sterilise them.
    b.  Now pick up an antibiotic disc from one of the petri dishes, and place it on one of your inoculated plates.
    c.  After placement on the agar, tap it once to make sure it is secure.

    Repeat steps a to c until you have placed this disc on a plate for each culture. Proceed to the next disc until five discs have been placed on a plate for each culture. Place the other five discs on the second plate, for a total of 10 discs per culture.
4.  When all discs are in place, put your eight plates into a 35°C incubator.

### *Examination of plates*

1.  Your plates must be examined after 16–18 hours of incubation. If you cannot examine them then, place them in a refrigerator until examination.
2.  Examine your plates for zones of inhibition. Measure these with a millimetre ruler across the disc as shown in Fig. 15.1 f. Record the diameter of the zone to the nearest whole millimetre. If only one side of the zone can be measured, multiply the number obtained by 2 to obtain a full zone of inhibition. If there is no zone (i.e. if growth occurs up to the edge of the disc), record a zero.

**Note**: You might see colonies within the zone of inhibition. These colonies consist of cells that are resistant to the antibiotic. Continue

## OBSERVATION AND RESULT

Record the diameter of the zones of inhibition

| Antibiotic | Disc code | Bacteria | | | |
|---|---|---|---|---|---|
| | | *Bacillus cereus* | *Escherichia coli* | *Pseudomonas aeuroginosa* | *Staphylococcus aureus* |
| | | | | | |
| | | | | | |
| | | | | | |
| | | | | | |
| | | | | | |
| | | | | | |
| | | | | | |

recording the zones of inhibition until you have all 40 measurements.

3. Now compare the zone of inhibition you obtained to the interpretive standards for these antibiotics. Record whether each organism is resistant, susceptible, or intermediate to the antibiotic.
4. Complete this exercise by recording for each type of bacteria the antibiotics they are susceptible to. These represent possible drugs of choice to treat infections by these bacteria.


## REVIEW QUESTIONS

1. Why is Mueller–Hinton agar used in antibiotic sensitivity test?

2. What agar should be used for testing fastidious organisms?

# 16

# SCREENING OF AMYLASE-PRODUCING ORGANISMS

## OBJECTIVES

To become familiar with
1.   screening amylase-producing organisms.
2.   the use of the enzyme amylase.
3.   the biochemical reaction catalysed by amylase.

## INTRODUCTION

Amylase is an exoenzyme that hydrolyses (cleaves) starch, a polysaccharide (a molecule which consists of eight or more monosaccharide molecules) into maltose, a disaccharide (double sugars, i.e. composed of two monosaccharide molecules) and some monosaccharides such as glucose. These disaccharides and monosaccharides enter into the cytoplasm of the bacterial cell through the semi permeable membrane and are thereby used by the endoenzymes. Starch is a complex carbohydrate (polysaccharide) composed of two constituents — amylose, a straight chain polymer of 200–300 glucose units, and amylopectin, a larger branched polymer with phosphate groups.

Amylase production is known in some bacteria while it is well-known in fungi. Amylases which are commercially produced from various Aspergilli are used in the initial steps of several food fermentation processes to convert starch to fermentable sugars. They are also used to partially predigest foods for young children, to clarify fruit juices and in the manufacture of corn and chocolate syrups.

## OBSERVATIONS

Examine the plates for the starch hydrolysis around the line of growth of each organism, i.e. the colour change of the medium.

## RESULTS

A typical positive starch hydrolysis reaction (i.e. a clear zone surrounding the microbial colonies) will be observed.

The ability to degrade starch is used as a criterion for the determination of amylase production by a microbe. In the laboratory it is tested by performing the starch test to determine the absence or presence of starch in the medium by using iodine solution as an indicator. Starch in the presence of iodine produces a dark-blue colouration of the medium, and a yellow zone around a colony in an otherwise blue medium indicates amylolytic activity.

This exercise deals with testing the hydrolysis of starch for the production of extracellular amylase by three test organisms, *Bacillus subtilis*, *Escherichia coli* and *Aspergillus niger* by inoculating these on starch agar medium.

## MATERIALS

### Culture

*Bacillus subtilis*
*Escherichia coli*
*Aspergillus niger*

### Media

Starch agar medium

### Miscellaneous

Gram's iodine solution
Sterile petri dishes
Dropper
Inoculating loop
Bunsen burner
Wax marking pencil

## PROCEDURE

1. Melt the starch agar medium, cool to 45°C and pour into the sterile petri dishes.
2. Allow it to solidify.
3. Label each of the starch agar plate with the name of the organism to be inoculated.

4.  Using the sterile technique, make a single streak-inoculation of
    each organism into the centre of its appropriately labelled plate.

5.  Incubate the bacterial inoculated plates for 48 hours at 37°C and
    fungal inoculated plates for 72–96 hours at 25°C in an inverted
    position.

6.  Flood the surface of the plates with iodine solution with a dropper
    for 30 seconds.

7.  Pour off the excess iodine solution.


## REVIEW  QUESTIONS

1.  What are the other substrates that can be used to screen amylase-
    producing organisms?

2.  Why is blue colour formed when iodine reacts with starch?

# 17

# SCREENING OF CELLULASE-PRODUCING MICROORGANISMS

## OBJECTIVE

To explain the following:

1. the biochemical reaction catalysed by cellulose
2. several sources of cellulose
3. secretion of cellulose by some bacteria and
4. commercial use of cellulose.

## INTRODUCTION

A prominent carbonaceous constituent of higher plants and probably the most abundant organic compound is cellulose. As a large part of the vegetation added to the soil is cellulose, the decomposition of the carbohydrate has a special significance in the biological cycle of carbon. Cellulose is a polysaccharide composed of glucose units in a long linear chain linked together by $\beta$-1, 4 glycosidic bonds. Cellulose is degraded by fungi, bacteria and actinomycetes by the secretion of an extraceullar enzyme called cellulase. It is a complex enzyme composed of at least three components viz. endoglucanase (endo-1, 4-$\beta$-D-glucanase (EC 3.2.1.4)), exoglucanase (1,4-$\beta$-D-glucancello-biohydrolase (EC 3.2.1.91)) and a $\beta$-glucosidase (EC 3.2.1.21). The cooperative action of these three enzymes is required for the complete hydrolysis of cellulose to glucose and this system has been completely characterised in *Trichoderma ressei*. The most extensively studied sources of cellulolytic enzymes are *Trichoderma, Phanerochaete, Cellulomonas* (aerobic bacteria) and *Clostridium thermocellum* (anaerobic bacteria).

## OBSERVATIONS

Observe the plates for the formation of a zone around the growth.

## RESULTS

A clear zone will be observed around the colonies of *B. cereus* and *T. viride* indicating degradation of CMC by the production of extracellular enzyme, i.e. cellulase (or CMCase).

This exercise deals with assaying the production of CMCase (endoglucanase), an extracellular enzyme, by two organisms, namely *Bacillus cereus* and *Trichoderma viride* by examining the inoculated Czapek-mineral salt supplemented with carboxymethyl cellulose agar plates, after 2–5 days of incubation.

## MATERIALS

### Culture

*Bacillus cereus*
*Trichoderma viride*

### Media

Modified Czapek-mineral salt medium

### Miscellaneous

Caboxymethyl cellulose (CMC)
Hexadecyltrimethyl ammonium bromide (1% solution)
Sterile petri plate (2)
Inoculating loop/needle
Glass rod
Bunsen burner
Wax marking pencil

## PROCEDURE

1. *Preparation of the medium*   Prepare 1 litre of the mineral Czapek-mineral salt agar medium whose constituents are as follows:

| | |
|---|---|
| Sodium nitrate ($NaNO_3$) | 2.0 g |
| Potassium phosphate ($K_2HPO_4$ or $KH_2PO_4$) | 1.0 g |
| Magnesium sulphate ($MgSO_4\ 7H_2O$) | 0.5 g |
| Potassium chloride (KCl) | 0.5 g |
| Carboxymethyl cellulose (CMC) | 5.0 g |
| Peptone | 2.0 g |
| Agar | 20.0 g |
| Distilled water | 1000 ml |

(a) Dissolve the agar in 400 ml of hot distilled water by adding in small amounts and stirring with a glass rod.
(b) Dissolve the magnesium sulphate, potassium chloride, peptone and sodium nitrate in 200 ml of water.
(c) Dissolve potassium phosphate in 100 ml of water.
(d) Dissolve CMC in 200 ml by heating and mix it in a blender.
(e) Mix all the solutions and make up the solution to 1000 ml volume.
(f) Adjust the pH of the medium to 6.5 with the addition of acid or alkali.
(g) Dispense the medium in 250-ml conical flasks, and sterilise it.

2. Pour the autoclaved medium into sterile petri plates.
3. Allow the medium to solidify.
4. Label the plates each with the organism that has to be inoculated.
5. Inoculate the appropriately labelled plates with the respective organism.
6. Incubate inoculated plates at 35°C in an inverted position for 2–5 days.
7. Flood the plates with 1% aqueous solution of hexadecyltrimethyl ammonium bromide.

**CAUTION**

- While preparing Czapek-mineral salt medium, phosphate should always be dissolved separately and be added to the mixture finally.

# 18

# IMMOBILISED WHOLE CELLS

---

## OBJECTIVE

To become familiar with immobilising yeast cells which contain many enzymes.

---

## INTRODUCTION

Immobilisation is a technique in which the cells are retained in a restricted space or surface of certain matrices but retain their catalytic activity. The methods available for immobilisation are classified into two broad types.

- Attachment on insoluble support
- Entrapment in a matrix

Because of the large size of the cells, entrapment is a widely used technique. Whole cells can be immobilised by entrapping them in spheres of alginate gel. This technique is commonly used in industry, and does not impair cell activity. There are several advantages in using immobilised cells; cells can be used for long periods of time, and the product is obtained uncontaminated with enzymes.

## MATERIALS

A length wide glass tubing
A one-holed rubber stopper
Screw clamp
Plastic gloves
Glass wool
Burette

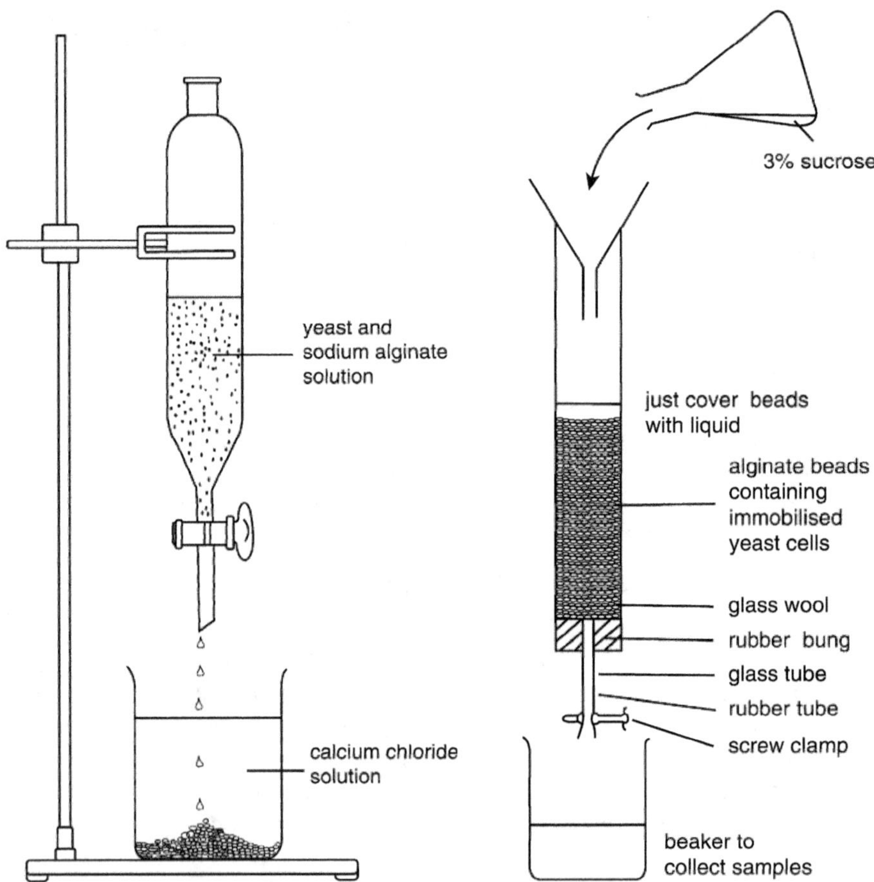


**Figure 18.1**   Formation of alginate beads.

**Figure 18.2**   A column of beads.

Clamp stand
Measuring cylinder
Filter funnel
Sodium alginate
Distilled water
Fresh baker's yeast
Calcium chloride
Benedict's reagent
3% Sucrose solution
Boiling tubes

## METHOD

1. Weigh out 4 g of sodium alginate.
2. Dissolve it in 100 ml of distilled water and stir thoroughly until it is completely mixed.
3. Weigh 5 g of fresh baker's yeast and blend it with 100 ml of distilled water to form a smooth paste in another 250-ml beaker.
4. Weigh 2.8 g of calcium chloride and dissolve in 200 ml of distilled water in a separate beaker.
5. Mix the yeast paste with the sodium alginate solution and stir well.
6. Support the burette in a clamp stand above the beaker of calcium chloride with the tip about 10 cm above the surface of the solution.
7. Fill the burette with the yeast and alginate mixture and allow drops of this to fall slowly into the calcium chloride solution (Fig. 18.1).
8. Stir the calcium chloride gently and observe the formation of beads of immobilised cells.
9. Allow the beads to stand for 20 minutes to cure and harden.
10. Set up the glass column as shown in Fig. 18.2 and insert some glass wool in the bottom of the column.
11. Carry out a Benedict's test for reducing sugar on a sample of 3% sucrose.
12. Strain the alginate beads through fine mesh, such as nylon tights, in filter funnel.
13. Rinse well with distilled water. Pack the glass column with alginate beads. Handle the beads gently to avoid damage.
14. Close the screw clamp and pour 3% sucrose into the glass column until the beads are just covered.
15. Collect approximately 1 ml of sample from the bottom of the column at 5-minute intervals and carry out a Bendict's test by adding 1 ml of Benedict's reagent and boiling the mixture.

## OBSERVATION AND RESULT

An orange colour indicates the presence of a reducing sugar.

## REVIEW QUESTIONS

1.  List the different types of solid matrices that can be used for immobilisation?

2.  List out some commercially important enzymes produced by yeast?

# 19

# MAKING SAUERKRAUT

## OBJECTIVE

To become acquainted with microbial production of sauerkraut.

## INTRODUCTION

For centuries humans have made use of fermentation as a means of preserving food for themselves and their animals. In many cases, the microorganisms that bring about the fermentation are naturally found on the substrate. Sauerkraut is a preserved food made from cabbage using the microorganisms naturally present in the cabbage to ferment the sugars in the cabbage to acids. Eventually the pH is so low that few other microorganisms can survive, thereby preserving the cabbage. In this experiment we are going to discuss about the preparation of sauerkraut and to investigate the changes that occur during fermentation.

## MATERIALS

Small solid drum-head cabbage
Chopping board and sharp knife
Sodium chloride
Clean glass jar
Cling film
Piece of polythene sheet (15 cm square)
Nutrient agar
Ethanol
Adhesive tape

## METHOD

1.  Strip the outer leaves from the cabbage, remove the core and discard the damaged parts, if any. Weigh the cabbage.
2.  Slice the cabbage as finely as you can using a sharp knife.
3.  Transfer the sliced cabbage to a large bowl and add 3% by weight of sodium chloride. Mix well and cover using a cling film. Leave for an hour, stirring at 10 minutes interval throughout this time.
4.  Pack the mixture into the gas jar and cover it using a polythene sheet with a weight on the top so that the cabbage is compressed and a layer of liquid forms above it. Note the colour, texture and smell of the cabbage.
5.  Measure the pH of the sample using a pH probe.
6.  Cover the jar of sauerkraut with a cling film, and label fully. Incubate at 22°C for about four weeks. Check the pH of the sample at weekly intervals. Also note the colour, texture and smell of the cabbage.

# 20

# MAKING GINGER BEER

## OBJECTIVE

To become familiar with microbial production of ginger beer.

## INTRODUCTION

A variety of fermented products can be made from microorganisms. Yeasts have been used for thousands of years to make alcoholic beverages such as wine and beer. In this practical we are going to discuss about the preparation of ginger beer.

## MATERIALS

Large white plastic bucket with lid
Large wooden or plastic spoon for stirring
Large saucepan
Cooker
Chopping board
Knife
2 lemons
1 kg granulated sugar
20 g of root ginger
25 g cream of tartar
25 g of dried yeast
Jug
Fine sieve
Plastic pop bottles with screw cap lids.

## OBSERVATION AND RESULTS

Leave the ginger beer in the bottle for two days and then drink it within the next two days.

## METHOD

1. Wash and cut the lemons into slices. Place the sliced lemons in the saucepan.
2. Chop up the root ginger and add it to the saucepan.
3. Add the cream of tartar and about half the sugar. Add about 3 l of water and boil the mixture stirring constantly.
4. Pour the contents of the saucepan into the fermentation bucket and add the rest of the sugar and 3.5 l of boiling water.
5. Stir well until all the sugars get dissolved, cover the bucket and leave it.
6. After about two days, the ginger beer is ready for bottling. Skim off the surface yeast froth, and strain the ginger beer through a fine sieve into a clean jug.
7. Using a funnel, transfer the beer from the jug into clean plastic pop bottles.
8. Fill the bottles, leaving an air gap at the top and then screw the lids.

# 21

# ETHANOIC ACID PRODUCTION

## OBJECTIVE

To generate a laboratory-scale mode of preparation of modern quick vinegar.

## INTRODUCTION

The discovery of vinegar must have been accidental, when ethanoic acid bacteria caused the souring of wine and beer. However vinegar came to be highly priced for both its flavouring and food preserving qualities.

## MATERIALS

A clean 1-litre plastic pop bottle
Non-adsorbent cotton wool
100 cm length plastic tubing
Aquarium type air diffusion block
Aquarium air pump
Fermentation lock with 10% sodium chlorate
Screw clip
Polystyrene packing material
10 ml of broth culture of *Acetobacter aceti*
pH probe
A can of beer

## METHOD

1.  Remove the screw cap from a 1-litre empty plastic pop bottle and rinse the bottle thoroughly with the sterilising solution provided.
2.  Attach the air diffusion block to the end of the clear plastic tubing and place the air diffusion block at the bottom of the pop bottle.
3.  Loosely pack the pop bottle with clean polystyrene packing material, broken piece of cork or large wood shavings.
4.  Pour the can of beer into the pop bottle. Keep a small sample in a test tube and determine the pH using a pH probe.
5.  Aseptically transfer the culture of *Acetobacter aceti* into the bottle.
6.  Plug the bottle loosely using non-absorbent cotton.
7.  Place the apparatus in a suitable place at a fairly warm ambient temperature.
8.  Ensure that the aquarium pump is above the level of the liquid in the bottle. Adjust the rate of bubbling using a screw clamp on the tube. Gentle aeration is sufficient.
9.  Remove a small sample every 24 hours and test the pH. Check the colour and smell of the liquor.

## RESULT

A quick vinegar is ready.

CAUTION

- You should not taste a food substance made in a laboratory.

## REVIEW QUESTIONS

1. List out the uses of vinegar?

2. List out the organisms that can produce vinegar?

# 22

# ISOLATION OF GENOMIC DNA

## OBJECTIVE

To become familiar with isolating genomic DNA from bacterial cells.

## INTRODUCTION

The preparation of high-quality chromosomal DNA is essential for many applications of genome analysis, including southern blotting, and the generation of recombinant DNA libraries.

## MATERIALS

### Culture

*Escherichia coli*

### Reagents

Extraction medium
NaCl                                      : 0.15 M
EDTA                                     : 0.1 M
SDS                                        : 25%
Sodium perchlorate               : 5 M
Ethanol                                  : 95%
Lysozyme                              : 10 µg/ml
Chloroform : isoamylalcohol  : (24 : 1)

## METHOD

1.  50 ml of the mid-log phase culture was taken and centrifuged at 8000 rpm for 10 minutes and approximately 2 g of the culture was taken.
2.  The harvested cells were mixed with 25 ml of the extraction medium and mixed well.
3.  1 ml of lysozyme at a concentration of (10 $\mu$g/ml) was added and incubated at room temperature for 30 minutes with occasional shaking.
4.  To the above mixture 20 ml of SDS solution was added and heated at 60°C for 10 minutes in a boiling water-bath.
5.  After incubating at 60°C the mixture was cooled using tap water. 6.5 ml of sodium perchlorate at a concentration of 5 M was added and mixed.
6.  Equal volume of chloroform : isoamylalcohol (24 : 1) was added and incubated at room temperature for 30 minutes by mixing slowly.
7.  The mixture was centrifuged at 12,000 rpm for 10 minutes.
8.  After centrifugation the aqueous phase was separated in a beaker.
9.  To the aqueous phase 95% of ethanol was added and the DNA was slowly spooled out.
10.  The spooled out DNA was redissolved in TE and stored at 4°C.
11.  The spooled DNA can be visualised by running it in agarose gel electrophoresis.

## REVIEW  QUESTIONS

1.  What is the purpose of

    (a) NaCl

    (b) EDTA

    (c) SDS

    (d) Sodium perchlorate

    (e) Lysozyme

    (f) chloroform : isoamyalcohol

# 23

# ISOLATION OF PLASMID DNA

## OBJECTIVE

To become familiar with isolating plasmid DNA from bacterial cells without any contamination.

## INTRODUCTION

Genetic engineering has introduced many new possibilities for employing microorganisms in the production of economically important substances. The key to genetic engineering is to introduce foreign genes into a microbial strain that will produce high quantities of desired products. Numerous vectors were used for transferring DNA from the donor to the recipient cell. These vectors range from plasmids of bacteria to viruses of mammalian cells. The biological molecules most commonly used for transferring genes into bacteria and yeast are the plasmids. Plasmids are small, double-stranded, closed, circular, self-replicating, extrachromosomal DNA molecules. Many methods have been developed to purify plasmids from bacteria. One method known as alkaline lysis method is described here.

### Alkaline lysis method

Alkaline lysis method involves lysis of cells with detergents in alkaline solution. Plasmid DNA extracted from *E.coli* using the alkaline lysis method was suitable for most molecular biological applications.

## MATERIALS

### Strain

E. coli/(pAMP) overnight culture.

### Media

LB medium with ampicillin

### Reagents

GTE Buffer
|  |  |
|---|---|
| Glucose | 50 mM |
| Tris | 25 mM |
| EDTA | 10 mM |

SDS/NaOH
|  |  |
|---|---|
| SDS | 1% |
| NaOH | 0.2 N |
| Potassium acetate (KOAc) | 3 M (pH 4.8) |
| Isopropanol | |
| 95% ethanol | |

Tris/EDTA (TE)                    (pH 8.0)
|  |  |
|---|---|
| Tris | 10 mM |
| EDTA | 1 mM |
| Chloroform : isoamylalcohol | 24 : 1 |
| DNase-free RNase | |

## METHOD

1. Transfer 1.0 ml of *E.coli*/pAMP overnight suspension into two eppendorff tubes.
2. Close the caps, and centrifuge at 10,000 rpm for 5 minutes to pellet the cells.
3. Pour the supernatant from the tubes into a beaker for later disinfection. Invert and wipe off the mouths of the tubes with a clean paper towel, to wick off as much of the remaining supernatant as possible.
4. Add 100 µl of ice-cold GTE to each tube. Resuspend the pellets by pipetting the solution in and out several times.

5. Add 200 µl of SDS/NaOH solution at room temperature to each tube. Close the caps, and mix the solutions by rapidly inverting the tubes about five times.
6. Place the tubes on ice for 5 minutes.
7. Add 150 µl of ice-cold KOAc solution to each tube. Close the caps and mix the solutions by rapidly inverting the tubes about five times.
8. Place the tubes on ice for 5 minutes.
9. Centrifuge the tubes at 12,000 rpm for 5 minutes to pellet the precipitate.
10. Transfer 400 µl of the supernatant from the tubes to a fresh tube and discard the old tubes containing the precipitate.
11. Add 400 µl of isopropanol to the supernatant. Close the caps and mix the solution by rapidly inverting the tubes about five times.
12. Centrifuge the tubes at 12,000 rpm for 5 minutes to pellet the precipitate.
13. Pour off the supernatant from the tubes and wick off as much as possible of the remaining alcohol on a paper towel.
14. Add 200 µl of 95 % ethanol to each tube, and close the caps. Flick the tubes several times to wash the pellets.
15. Centrifuge the tubes at 12,000 rpm for 5 minutes to pellet the precipitate.
16. Pour off the supernatant from the tubes and wick off as much as possible of the remaining alcohol on a paper towel.
17. Add 5 µl of RNase at 1 mg/ml concentration (DNase free) and leave it in a water bath at 37°C for 15 minutes.
18. Arrest the reaction by adding an equal volume of phenol i.e. 100 µl and spin at 10,000 rpm for 10 minutes.
19. Without disturbing the lower phase take only the upper aqueous phase in a fresh eppendorf tube using the tip.
20. Add 0.1 ml of chloroform : isoamylalcohol (24 : 1).
21. Mix and spin at 10,000 rpm for 5 minutes.
22. Take the supernatant without disturbing the lower phase and transfer it to a fresh eppendorff tube, add 2.5 volumes of ice-cold ethanol in the presence of 0.25 M ammonium acetate.
23. Mix, leave it at –20°C overnight.
24. Spin at 10,000 rpm for 15 minutes. Wash the pellet with 70% ethanol.
25. Dry the nucleic acid by exposing it to a stream of warm air or air-dry the pellets.

26. Add 15 µl of TE buffer to the tubes. Resuspend the pellets by scraping them with a pipette tip and vigorously pipetting in and out. Check that all the DNA is dissolved and that no particles remain in the tip.
27. Freeze the DNA TE solution at –20°C until use.

## REVIEW QUESTIONS

1. Define CCC.

2. What is the purpose of

   (a) GTE

   (b) SDS/NaOH

   (c) KOAc

   (d) Isopropanol

# 24

# RAPID ISOLATION OF PLASMID USING TELT

## OBJECTIVE

To become familiar with rapid isolation of plasmid DNA from bacterial cells.

## INTRODUCTION

This is a rapid technique developed to isolate small amounts of DNA which is not very pure but still suitable for some purposes.

## MATERIALS

### Culture and reagents

Host strain with plasmid
LB broth
Microfuge tubes
Gloves
TELT solution
Tris HCl 50 mM (pH 8.0)
EDTA 62.5 mM
Lithium chloride 2.5 M
Triton X – 100  4% (V/V)
Phenol : Chloroform 1 : 1 (V/V)
RNase A
Ethanol 95%
TE Buffer
Tris  10 mM
EDTA  1 mM  (pH 8.0)

## METHOD

1.  The plasmid-containing cultures were treated according to their specific selection characteristics.
2.  Transfer 1.0 ml of the culture to a clean eppendorff and centrifuge at 10,000 rpm for 5 minutes to pellet the cells.
3.  Pour off the supernatant and resuspend the cells in 100 μl of TELT solution.
4.  Add 100 μl of 1 : 1 phenol : chloroform and vortex vigorously.
5.  Centrifuge at 12,000 rpm for 2 minutes and transfer the aqueous layer into a fresh eppendorff tube containing an equal volume of 95% ethanol.
6.  Allow to stand for 10 minutes and centrifuge at 12,000 rpm for 10 minutes.
7.  Pour off the supernatant, dry the pellet and dissolve the pellet in 25 μl of TE buffer.
8.  Freeze the DNA at −20°C until use.

# 25

# VISUALISATION OF DNA IN AGAROSE GEL

## OBJECTIVE

To become familiar with electrophoretic technique and to visualise DNA bands in cells.

## INTRODUCTION

Agarose is a linear polysaccharide made up of a basic repeat unit of agarobiose, which comprises alternating units of D-galactose and 3,6-anhydrogalactose. Agarose gel electrophoresis is a technique used to resolve DNA fragments on the basis of their molecular weight. Smaller fragments migrate faster than larger ones. The distance migrated on the gel varies inversely with the logarithm of the molecular weight. Calibrating the gel using known size standards and comparing the migration distance of the unknown fragment, you can determine the size of the fragment.

## MATERIALS

TAE buffer (50 X) in 100 ml
  Tris base - 24.2 g
  Glacial acetic acid - 5.71 ml
  0.5 M EDTA - 10.0 ml
Agarose
Ethidium bromide - 10 mg/ml

## Results

**Name:**________________

Gel preparation: ________%gel    Agarose powder:________g

Sample volume: ________µl      Stain: ____________________

Well data

| Well number | 1 | 2 | 3 | 4 | 5 | 6 | 7 |
|---|---|---|---|---|---|---|---|
| Contents | ____ | ____ | ____ | ____ | ____ | ____ | ____ |
| Number of bands | ____ | ____ | ____ | ____ | ____ | ____ | ____ |
| Migration (mm) | ____ | ____ | ____ | ____ | ____ | ____ | ____ |

Interpretation

Tracking dye / Loading buffer (5 ml)
   Bromophenol blue - 12.5 mg
   Xylene cyanal - 12.5 mg
   Sucrose - 2.0 g

## METHOD

1. Prepare 0.8% agarose by adding 0.8 g of agarose to 100 ml of 1X TAE buffer.
2. Warm the mixture in a flask at 100°C in a water bath until the agarose is dissolved.
3. Cool the agarose solution until its temperature reaches 50°C.
4. Add 10 µl of ethidium bromide to the agarose. Swirl the flask to avoid air bubbles while mixing.
5. The gel tray was tapped on the two sides and to prevent liquid agarose from leaking out of the tapped ends of the gel tray seal the sticky side with a thin line of cooled agarose solution.
6. With gel tray on a level surface place the comb into the notched end of the tray and pour the contents of the flask into the tray and let the solution set.
7. After the gel has solidified remove the comb and tape from the gel tray. Place the tray on a raised platform within the apparatus in such a way that the sample wells are close to the negative black electrode.
8. Pour 1X TAE buffer into the gel tank in such a way that the gel is fully immersed in the buffer. Remove bubbles from the wells using a pipette.
9. To a 1.5-ml microcentrifuge tube, add 5 µl of 10X gel loading buffer and 45 µl of redissolved DNA. Mix by tapping the tube.
10. Load up to 20 µl of the solution mixture to fill a well in the agarose gel.
11. Turn the power supply at 50 volts and stop when the first dye front is 80% down.
12. After the electrophoresis is completed, turn off the power and disconnect the leads and remove the cover.
13. Remove the gel from the platform and carefully transfer it to the UV transilluminator for visualisation.
14. The gel can be photographed to obtain a permanent record of the band pattern in the gel.
15. After visualisation carefully remove the gel from the illuminator using two spatulas and safely discard it.

## REVIEW QUESTIONS

1. Define

   a. Electrophoresis

   b. Agarose

2. What is the use of a buffer in electrophoresis? Why is TAW advantageous over TBE?

3. How will you calculate the molecular weight of the DNA?

# 26

# QUANTITATION OF DNA BY SPECTROPHOTOMETRY

## OBJECTIVE

To construct a standard graph using a known DNA sample and to determine the quantity of the unknown DNA sample.

## INTRODUCTION

Researchers should judge the quantity and quality of DNA for use in cloning, transfection, southern blotting, restriction enzyme analysis, sequencing, polymerase chain reaction and transformation. There are several methods available to determine the quantity and concentration of DNA in a sample. These include adsorption spectrophotometry, fluorescence spectrophotometry and agarose gel electrophoresis. Although fluorescence spectrophotometry is the most sensitive one in detecting DNA at concentrations of 0.01 to 15 µg/ml, absorption spectrophotometry is still the standard method. It detects 1 to 50 µg of DNA per ml. Adsorption spectrophotometry is the most reliable method for assessing both the quantity and purity of DNA. However this method requires large quantities of DNA to obtain accurate readings and cannot distinguish between DNA and RNA.

## MATERIALS

### Reagents

DNA standard solution
Unknown DNA sample
10X TNE (Tris-sodium chloride-EDTA buffer)

## OBSERVATION AND RESULT

Using the standard curve determine the concentration of the unknown DNA sample.

Estimate the purity of the DNA by determining the $A_{260}/A_{280}$.

## Equipment

UV-spectrophotometer

## Miscellaneous

Pipettes
Test tubes
Tissue paper

## METHOD

1. Prepare dilutions of known, standard DNA in 1X TNE for a standard curve. Choose a final volume appropriate for use in spectrophotometer.
2. Measure the absorbance of the diluted DNA with TNE as blank and record the results as given in the table.

| Concentration of DNA (g/ml) | Absorbance | |
|---|---|---|
| | 260 nm | 280 nm |
| 1000.0 | | |
| 100.0 | | |
| 10.0 | | |
| 1.0 | | |

1. Construct the standard graph using the data for $A_{260}$. Plot the absorbance on the Y-axis and DNA concentration on the X-axis.
2. Place the unknown DNA sample in the spectrophotometer and record the absorbance.
3. Using the absorbance, calculate the concentration of the unknown sample by extrapolating from your standard curve.
4. Determine the $A_{260}/A_{280}$ ratio to estimate the purity of the sample. If the ratio lies between 1.8 and 2.0 then the DNA sample is said to be free from proteins.

# 27

# RESTRICTION ENZYME DIGESTION AND SIZE DETERMINATION OF DNA

## OBJECTIVES

1. To become familiar with the use of restriction enzymes to digest DNA.
2. To construct a standard graph using the marker DNA.
3. To determine the size of an unknown DNA.

## INTRODUCTION

Restriction enzymes are bacterial enzymes that cleave double-stranded DNA. Type II restriction endonucleases are highly site-specific for DNA cleavage and thus act as extremely useful tools in molecular biology. Type II Restriction endonucleases recognise and cut within the palindromic sites. An example is the *EcoRI* which recognises the sequence 5'- G$^\downarrow$AATTC. Since the cut site is between the G and A bases, this will lead to fragments having overlapping sticky ends.

In this experiment the lambda DNA will be incubated with the restriction enzymes *HindIII,* and *EcoRI.* Subsequent gel electrophoresis of the individual reaction mixtures will illustrate the different fragmentation patterns produced by the enzymes. The size of the particular piece of DNA can be obtained by procedures applied for proteins. A molecular weight standard is run on a separate lane on the same gel. The standard fragments are all of known size and can be used to construct a curve relating to the molecular weight versus migration distance in the gel. The molecular weight of the unknown DNA can be determined from the standard graph by measuring the migration distance in the gel. Agarose gels are used instead with adjustments in agarose percentage in order to accommodate the possible range in size of DNA pieces that might be encountered. The table below shows some typical agarose gel percentages appropriate for various DNA sizes.

| Agarose (%) | Effective range of resolution of linear DNA fragments (kb) |
|---|---|
| 0.5 | 30–1 |
| 0.7 | 12–0.8 |
| 1.0 | 10–0.5 |
| 1.2 | 7–0.4 |
| 1.5 | 3–0.2 |

## MATERIALS

Lambda DNA – 0.5 mg/ml in TE buffer
*Eco RI*
*Hind III*
*Bam HI*
10X Enzyme buffer
Microfuge tubes
Tracking dye
Ethidium bromide (10 mg /ml)
Agarose
1X TBE Buffer
Electrophoresis unit
Water bath
Transilluminator

## METHOD

1.  Set up a restriction digestion as follows using three 1.5-ml microfuge tubes.

|                              | Tube I | Tube II | Tube III |
|------------------------------|--------|---------|----------|
| Lambda DNA                   | 2 µl   | 4 µl    | 4 µl     |
| 10X *Eco RI* Enzyme buffer   | –      | 1 µl    | –        |
| 10X *Hind III* Enzyme buffer | –      | –       | 1 µl     |
| Distilled water              | 8 µl   | 4 µl    | 4 µl     |
| *Eco RI*                     | –      | 1 µl    | –        |
| *Hind III*                   | –      | –       | 1 µl     |

2. Incubate at 37°C for one hour.
3. Denature proteins by heating in a 75°C water-bath for 5 minutes.
4. Prepare a 0.75% agarose in 1X TBE. Boil the mixture until the agarose dissolves. Cool it to 50°C and then transfer 10 µl of ethidium bromide and mix well.
5. Place the well comb in a position and pour enough of the warm agarose to form a gel of the desired thickness.
6. Add 2 µl of tracking dye to each digest and load the samples in the gel and electrophorese at 60 V for hours or until the dye travels 0% of the distance in the gel.

## RESULTS

1. Determine the migration distance in millimetres for lambda DNA and each *Hind III* fragment by measuring the distance from the bottom of the well to the middle of the fragment.
2. Using a semilog graph paper, construct a standard curve by plotting the size on the log axis Vs migration distance on the linear axis.
3. Choose one or more appropriate fragments from the *Eco RI* lane and measure their migration distance.
4. Locate the distance on the linear axis of the standard curve and find the point where that value intersects the curve.

> The size in base pairs of the fragments from digestion
> of lambda DNA are
>
> *Hind III* – 23,130; 9,416; 4,361; 2,322; 2,027; 564; 125
>
> *Eco RI*: 21,226; 7,421; 5,804; 5,643; 4,878; 3,530.

## REVIEW QUESTIONS

1.  Define

    (a) Endonucleases

    (b) Exonucleases

# 28

# TRANSFORMATION IN *E. COLI*

## OBJECTIVE

To perform transformation technique and to screen the recombinants.

## INTRODUCTION

Transformation is a process by which a cell takes up a naked DNA from the surrounding medium and incorporates it to acquire an altered genotype that is heritable. Natural transformation serves to add genetic variability to the population. Cells that have the ability to undergo natural transformation are said to be naturally competent. In order for a host cell to undergo transformation, it must be competent. Other bacteria such as *E.coli* may be artificially induced to become competent and then can be transformed.

Competence can be induced by treating the cells with cations like $CaCl_2$ but the transformation efficiency of the *E.coli* that is transformed using $CaCl_2$ ranges from $10^6$ to $10^8$ transformants per microgram of plasmid DNA. When high efficiency is desired electroporation is the method of choice. Electroporation uses a high-voltage charge to make transient holes in the bacterial membrane through which the plasmid may move.

Transformation can be carried out in four steps.

1.  **Pre-incubation** The cells are suspended in a solution of cations incubated at 0°C. The cations form a complex with the negatively charged phosphates of lipids in the *E.coli* cell membrane. The low temperature congeals the cell membrane, stabilising the distribution of charged phosphates allowing them to be more effectively shielded by the cations.

## OBSERVATION AND RESULTS

Emergence of colonies in ampicillin-selective plates shows that transformation has occurred. Confirm the presence of plasmid in the ampicillin-resistant transformants by isolating the plasmid from the colonies.

2. **Incubation**   The DNA is added, and the cell suspension is further incubated at 0°C. The cations are said to neutralise negatively charged phosphates in the DNA and the cell membrane.

3. **Heat shock**   The cell suspension along with the DNA is briefly incubated at 42°C and then returned to 0°C. The rapid temperature change creates a thermal imbalance on either side of the *E.coli* cell membrane, which is said to create a draft that sweeps the plasmid into the cell.

4. **Recovery**   LB broth is added to the suspension and incubated at 37°C before plating on a selective media. Transformed cells recover from the treatment.

## MATERIALS

*E.coli* ($DH_5 \alpha$)
pBR 322
LB agar with ampicillin (50 mg/l)
Ice-cold $CaCl_2$ solution (100 mM)
$MgCl_2$ (100 mM)
Ice cubes
Micropipettes

## METHOD

1. Inoculate 1 ml of the overnight culture into 100 ml of LB broth and incubate at 37°C in a rotary shaker.

2. Allow it to grow till the culture reaches 0.3 to 0.6 O.D at 600 nm.

3. Transfer 25 ml of the culture to a centrifuge tube and harvest the cells by centrifuging at 10,000 rpm for 10 minutes and discarding the supernatant.

4. Suspend the pellets in 10 ml of ice-cold magnesium chloride solution.

5. Centrifuge at 10,000 rpm at 4°C for 5 minutes and discard the supernatant.

6. Suspend the pellets in 2 ml of ice-cold calcium chloride solution and place the tubes in ice for 30 minutes.

7. Suspend the pellets in 800 ml of ice-cold magnesium chloride solution. Dispense 200 ml into pre-chilled, sterile tubes. Freeze immediately at – 70°C.

8. For transformation take the competent cells from –70°C and thaw it over ice. Transfer 100 µl of the competent cells to a pre-chilled

eppendorff tube, add 1 µg of plasmid DNA and leave it in ice for 10 minutes.

9. Heat shock cells by placing the tubes in 42°C water bath for 2 minutes and immediately return the tubes to ice and allow them to stand in ice for 1 minute.

10. Place the tubes in the test tube rack at room temperature.

11. Make the volume to 1 ml with sterile LB medium and gently tap the tubes to mix.

12. Allow the bacteria to recover from the heat shock by placing at room temperature for about 30 to 45 minutes.

13. Transfer 0.1 ml of the culture to LB agar plates containing ampicillin and spread the sample over the plates using a sterile spreader.

14. Incubate the plates at 37°C for 24–48 hours. Observe and count the number of transformants.

## REVIEW  QUESTIONS

1. Do you think that cells that grow on ampicillin plate are also resistant to tetracycline? Why or why not?

2. Why are most antibiotics safe case to humans even though they can be harmful to bacteria?

3. How does ampicillin kill ampicillin-sensitive bacteria? How do ampicillin-resistant bacteria avoid being killed by ampicillin?

4. Why is having an antibiotic resistance gene on a plasmid more beneficial to the bacterium than having the gene on the bacterial chromosome?

# 29

# NTG — MUTAGENESIS OF *E. COLI*

## OBJECTIVE

To bring about mutation in bacteria using chemical mutagens.

## INTRODUCTION

Mutation is a natural metabolic process that occurs spontaneously at a low frequency. We do not know for certain how spontaneous mutations arise in nature. It has been observed that the process of DNA replication is somewhat error-prone. While most replicative errors are corrected then and there or soon afterwards, a few of them escape and are manifested as spontaneous mutations.

Treatment of cells with some physical and chemical agents called "mutagens" increases the frequency of mutations. Such agents are useful in the laboratory to generate mutations which in turn has contributed to the development of genetics.

In this experiment a potent mutagen, namely N-methyl N-nitrosoguanidine also known as NTG is used to generate mutations.

In this experiment, streptomycin-sensitive strains of *E.coli* were taken and treated with mutagens and specifically streptomycin-resistant mutants were screened.

## MATERIALS

*E. coli* KL 16/96 rif$^r$, kan$^r$, str$^s$
LB media

## OBSERVATION AND RESULTS

Frequency of spontaneous mutants = TSM/TVC
Percentage of spontaneous mutants = TSM/TVC × 100
Frequency of induced mutants = TIM/TVC
Percentage of induced mutants = TIM/TVC × 100
Frequency of streptomycin-resistant mutants = IM/TIM
Percentage of streptomycin-resistant mutants = IM/TIM × 100

LB selective media (media with streptomycin 100 µg/ml)
Stock solutions
Saline (0.85%)
NTG stock (10 mg/ml)
Streptomycin stock (1000 mg/ml)

## METHOD

1. Dilute the overnight culture of *E.coli* KL16/96 LB. Agitate at 37°C for 3 hrs.
2. Transfer 1 ml of the overnight culture into 3 sterile eppendorff tubes and harvest the cells by centrifuging at 6,000 rpm for 10 minutes.
3. Wash the cell pellet twice using sterile saline.
4. Resuspend the pellet in sterile saline.
5. Take the first eppendorff tube serially dilute and plate it on sterile air-dried LB plates to determine the total viable count.
6. Spread plate the undiluted inoculum from the second eppendorf on LB selective plates to determine the total spontaneous mutants.
7. To the third tube add NTG stock (50 mg/ml) and incubate it at 37°C for 20 minutes.
8. Centrifuge the NTG-treated cells at 6,000 rpm for 10 minutes and collect the cell pellet. Wash the cell pellet thrice with sterile saline.
9. Resuspend in 10 ml of sterile LB broth and incubate at 37°C overnight.
10. Take 1 ml of the overnight culture, serially dilute it with sterile saline and plate it on an air-dried LB medium to detect the total induced mutants and that on the LB selective media to detect the streptomycin mutants (induced mutants).

## REVIEW  QUESTION

1.   What are the various chemical mutagens available?

2.   Define mutation?

3.   What are the various methods of inducing mutation?

# 30

# AMPICILLIN SELECTION OF AUXOTROPHS

## OBJECTIVE

To enrich auxotrophs using ampicillin and to calculate the auxotroph frequency after mutagenesis.

## INTRODUCTION

Ampicillin, a derivative of penicillin, has the capacity to kill growing bacterial cells by interfering with the terminal reaction in cell wall synthesis. This provides a selection technique for enriching auxotrophs from a population. In a minimal medium, among the population, prototrophs alone can grow and hence the addition of ampicillin lyses the growing cells. As auxotrophs cannot grow in minimal media, they escape during ampicillin treatment. The main disadvantage of ampicillin treatment is that sometimes the extensive cell lysis of the prototroph cell leads to the cross feeding of auxotrophs thereby killing them. This problem can be circumvented if a hypertonic medium with glucose or sucrose is used. This converts growing cells into protoplasts and hence they do not lyse.

## MATERIALS

Petri plates
Pipettes
Eppendorf tubes
Test tubes
L-rod
Toothpicks

## OBSERVATION AND RESULT

$$\text{Auxotroph frequency} = \frac{\text{Number of auxotrophs}}{\text{Number patched}}$$

### Culture

MNNG-treated culture of *E.coli* KL 16/96

### Media used

LB medium
Glucose minimal medium

### Stock solutions

Ampicillin (1gm/100 ml)
Saline (0.85%)
Glucose (40%)

## METHOD

1.  Dilute the overnight culture of NTG treated *E.coli* in 5 ml of glucose minimal medium.
2.  Aerate at 37°C until it reaches early log phase (90–120 minutes).
3.  Add ampicillin 100 µg/ml, which kills the growing cells.
4.  Centrifuge at 10,000 rpm for 10 minutes. The cells that were collected were mainly auxotrophs and some were prototrophs that escaped from the ampicillin treatment.
5.  Resuspend the pellets in LB medium and incubate overnight.
6.  Dilute the overnight culture using sterile saline and plate the sample on LB plates by spread plating technique.
7.  Patch atleast 40–50 CFU on LB + minimal medium with sterile toothpicks.
8.  Incubate the plates at 37°C and calculate the auxotroph frequency.

## REVIEW QUESTIONS

1.  Why is ampicillin used instead of penicillin?

2.  Define (a) auxotroph (b) prototroph?

# 31

# MUTAGENESIS IN BACTERIA: THE AMES TEST

## OBJECTIVE

To test different chemicals for carcinogenicity.

## INTRODUCTION

An animal or plant cell becomes cancerous when it accumulates mutations that lead to unregulated cell division, chromosomal instability, and/or the inability to undergo normal cell death (apoptosis). Therefore, any natural or synthetic agent that damages DNA is a potential carcinogen. In 1971, Dr. Bruce Ames developed a rapid method for identifying mutagens — and so, potential carcinogens — using a special strain *Salmonella enterica* (formerly *S. typhimurium*). The strain has two features that make it ideal as a sensor for mutagens. Firstly, it lacks DNA repair enzymes so that mistakes in DNA synthesis are not corrected. Secondly, it carries a point mutation that renders it a histidine auxotroph (his$^-$). It is unable to synthesise this amino acid from the ingredients in its culture medium. In the presence of a mutagen, reversions or back mutations to the his$^-$ phenotype occur at a high rate, and the revertants are easily identified.

In the Ames test, the auxotrophic strain is exposed to a test chemical and cultured on a nutrient medium containing only a small amount of histidine. The his$^-$ cells can survive until their histidine is used up. Cells that have reverted to the his$^+$ phenotype continue to grow even in the absence of exogenous histidine.

The number of colonies on the test plate is therefore proportional to the efficiency of the mutagen.

This bacteria-based mutagenesis test provides a fast, inexpensive way to identify potential carcinogens. It is important to note, however, that some substances that cause cancer in laboratory animals are not mutagenic in the Ames test, and some substances identified as mutagens in the Ames test do not appear to cause cancer. Some chemicals (called pro-mutagens) are not mutagenic unless they are converted to more active derivatives by liver enzymes. For example, benzo[a]pyrene is not mutagenic, but it is converted by liver enzymes to diolepoxides which are potent mutagens and carcinogens. Therefore, to test for pro-mutagens an extract of rat liver enzymes is usually included in the Ames test.

Since *Salmonella* is pathogenic in humans, we will be using a harmless strain of *E. coli* that is auxotrophic with respect to histidine (and thiamine) as our mutagen-sensor strain. Although this strain is not optimised for mutagenesis (it is capable of DNA repair), the principle of the test is the same. In addition, we will not include liver enzymes in the test, so we will not be testing for pro-mutagens.

## MATERIALS

### Cultures

Overnight culture of *E. coli* strain AB 3612 in nutrient broth or *S. typhimurium*, Ames test strain

### Media

10 minimal medium agar plates (per group)
40 plates

### Reagents

10X thiamine solution (20 mg/ml)
Sterile distilled water (for water-soluble solids to be tested)
Chloroform (for water-insoluble solids to be tested)
70% ethanol in a shallow dish

Test substances provided in the laboratory (such as diethyl sulphate, 4 nitro-o-phenylenediamine or sodium nitrite) and those supplied by students (such as household products). The effects of UV radiation can also be tested if a UV lamp is available, along with UV-safe goggles and gloves.

## Equipment

37°C incubator with shaker platform
Bunsen burner

## Miscellaneous

Sterile Pasteur pipettes/bulb or
Transfer pipettes
Microfuge tubes (8)
1.0 ml serological pipette
Micropipettes/tips (100–1,000 µl)
Spreader
Sterile forceps
Sterile filter paper discs (0.75 cm diameter)
Laboratory marker

## PROCEDURE

1. Obtain ten minimal medium agar plates, four plates for each substance you are testing (2 substances) and two control plates. Pipette 0.1 ml of 10X thiamine solution onto each of the minimal medium agar plates. Distribute the liquid as evenly as possible with a sterile spreader.
2. Once the plates have been dried, pipette 0.1 ml of the overnight culture of *E. coli* strain AB 3612 onto each plate with a sterilised spreader. Spread the cells as evenly as you can.
3. While the plates are drying, prepare two substances that you wish to test for mutagenicity. If the material is a solid, weigh out 1 mg using an analytical balance, place it in a microfuge tube, and dissolve it in 1 ml of sterile, distilled water.

**Note**: If the substance does not dissolve in water, weigh out another milligram, and dissolve it in 1 ml of chloroform. If the substance is a liquid, record its concentration if known.

# RESULTS

Complete the following table

Test substance
Substance description
Concentration (mg/ml)

| Sample number | Sample dilution | Total material tested | Number and distribution of any colonies | Mutagenic at this level |
|---|---|---|---|---|
|  |  |  |  |  |
|  |  |  |  |  |
|  |  |  |  |  |
|  |  |  |  |  |
|  |  |  |  |  |

4. Prepare dilutions of both liquids. For each substance to be tested, label three microfuge tubes with the name of the test substance, and label them 2,3 and 4 (tube 1 is the original, undiluted sample). Pipette 1.0 ml of the appropriate diluent (chloroform or sterile water) into the tubes numbered 2,3, and 4, if you use chloroform, keep the tubes capped. Then, using a micropipette, transfer 1 µl of the undiluted liquid sample into tube 2, and mix well. Using the same tip, transfer 1 µl of sample from tube 2 to tube 3, and mix well. Using the same tip, transfer 1 µl of sample from tube 3 to tube 4 and mix well. Repeat this series of dilutions on the second liquid.

5. Label the plates the same way you labelled the microfuge tubes (1–4 and substance name). Label the two remaining plates "dry disc control" and "solvent control".

6. Using sterile forceps, place a sterile filter paper disc at the centre of each plate.

7. Using sterile Pasteur pipette or transfer pipette, add 1 drop of a liquid sample to the centre of the filter paper disc on the corresponding plate. The filter paper should be saturated but not dripping wet. If needed, add additional sample, drop by drop, until the paper is saturated. Count the number of drops you use.

8. To the "dry disc control" plate, add no liquid. To the "solvent control" plate, add either sterile water or chloroform, drop-wise, as in step 7. If you used both solvents, choose just one, but be sure that someone else in the class performs the other solvent control.

9. Incubate the plates at 37°C, in an inverted position. Be sure that the disc continues to adhere to the agar. Incubate the plates for 2 days (the plates will then be stored in the refrigerator).

10. Examine your plates, and record the results in your laboratory report.


## REVIEW QUESTIONS

1. An overnight culture is expected to be at the stationary phase of growth and at a density of about 109 cells/ml. Given this approximately how many cells did you plate initially?

2. If you detected revertant colonies on any of the plates, select one plate
   and do the following
   (a) Calculate the approximate surface area (cm$^2$) of the region where
       the colonies appear.

   (b) What is the "mutagenesis efficiency" of this substance, expressed
       as the number of revertant colonies per total number of cells
       plated.

# 32

# ISOLATION OF BACTERIOPHAGE FROM SEWAGE AND DETERMINATION OF PHAGE TITER

---

## OBJECTIVES

1. To isolate bacteriophages from sewage.
2. To determine phase titer.

---

## INTRODUCTION

Virtually any type of cell is susceptible to viral infection. Viruses cause disease in plants and animals, and can also infect prokaryotes and unicellular eukaryotes. Viruses that infect prokaryotes are known as bacteriophages or phages, because when they were first discovered, they appeared to eat bacterial cells, generating a clearing, or plaque, on a lawn of susceptible bacteria. In reality, the bacteria are killed by lysis as newly produced phages are released from the damaged cells.

Like all viruses, bacteriophages consist of nucleic acid (RNA or DNA) surrounded by a protein coat, or capsid. Unlike some plant and animal viruses, bacteriophages are not enveloped. Some phages have elaborate structures for attaching to the bacterial surface and injecting nucleic acid into the cytoplasm.

Most bacteriophages are lytic; that is, each infection event leads to the production of new virions and the death of the cell by lysis. Some bacteriophages — most notably bacteriophage lambda ($\lambda$) — are categorised as temperate. Sometimes $\lambda$ DNA is integrated into the bacterial chromosome, with its genes largely silent. The infected cell

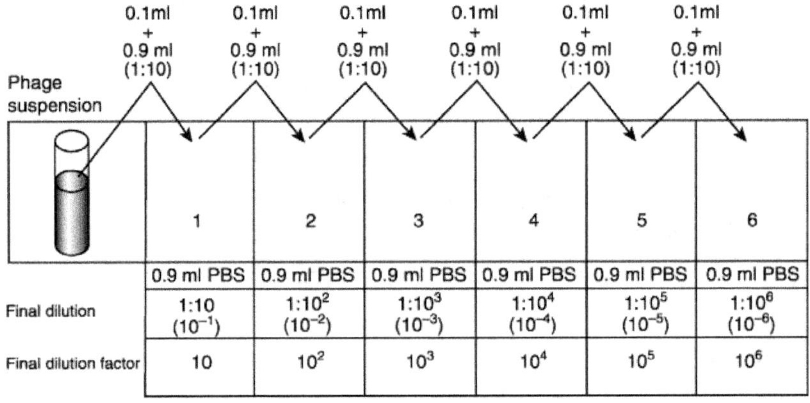



**Figure 32.1** Serial dilutions of bacteriophage suspension.

survives as a lysogen. In some λ infections, the DNA remains independent of the host chromosome, and is replicated many times over. Its genes are expressed at high levels, and newly assembled phages are released. The "choice" between a lysogenic, nonproductive infection and a lytic, productive infection depends on environmental conditions. For example, UV exposure can cause a λ infection to switch from lysogenic to lytic.

In this exercise, we will focus on the bacteriophage of coliform bacteria. Coliform bacteria are relatively harmless microorganisms that live in large numbers in the intestines of mammals, where they aid in the digestion of food. *Escherichia coli* is a common faecal coliform bacterium. The presence of faecal coliform bacteria in water indicates that it has been contaminated with human or other animal faeces, and that a potential health risk exists for those who use the water. Raw, untreated sewage contains large numbers of *E. coli*. Therefore, we will use raw sewage as a source of the bacteriophage that infects *E. coli*.

In this exercise, you have the opportunity to: (a) amplify (increase the numbers of) phages in the sewage sample by allowing them to infect and reproduce within fresh *E. coli*, (b) collect the phages from the culture by centrifugation and filtration, and (c) detect and titer the amplified, isolated phages using a plaque assay. The assay is based on the fact that each plaque on a lawn of bacteria, although it contains $10^6$ to $10^7$ virions along with bacterial debris, represents a single infecting phage that entered one cell at the start of the culture. The infection then "spread" as the viruses reproduced and cells lysed, eventually forming a visible plaque. The titer of a phage suspension, therefore, is determined by counting the number of plaques that form from a given volume of suspension. Phage titer is expressed as plaque forming units (PFU) per millilitre (ml).


## MATERIALS

### Cultures

Overnight culture of *E.coli*
40 ml of raw sewage

### Media

Nutrient broth
10X strength nutrient broth

Warmed nutrient agar plates (6, 100 × 15 mm plates)
Tubes containing 3 ml each of warm, top
    agarose (one per plate)
7.5 g agarose/litre nutrient broth
    (molten, cooled to 45°C)

## Reagents

Phosphate-buffered saline (PBS)

## Equipment

37°C incubator with shaker platform
Water bath at 37°C
Water bath at 45°C

## Miscellaneous

5-ml pipettes
15-ml conical centrifuge tube
Tube for collection and storage of phage filtrate
Sterile 0.45-mm syringe tip filter
10-ml syringe without needle
1.5-ml microfuge tubes for preparing dilutions
1.0-ml serological pipette
micropipette/tips (100–1,000 ml)
Laboratory marker

## PROCEDURE

Raw sewage was collected from a local sewage treatment plant prior to the day of the experiment and 50 ml of 1X nutrient broth was inoculated with *E. coli* for overnight growth at 37°C with shaking.

*Amplification of bacterial viruses*

1.  Pipette 5 ml of 10X nutrient broth into the flask containing 40 ml of raw sewage.
2.  Inoculate the sewage in the flask with 5 ml of an overnight culture of *E. coli*.

3. Inoculate a separate flask containing 45 ml of 1X nutrient broth with 5 ml of an overnight culture of *E. coli*.
4. Incubate both cultures at 37°C in a shaker for 24 hours.

*Bacteriophage Isolation and Plating*   Prior to today's lab, 2 ml of 1X nutrient broth was inoculated with *E. coli* for overnight growth at 37°C with shaking. Earlier today, 100 ml of 1X nutrient broth was inoculated with a small volume of the overnight. This was done to obtain a culture in log growth by class time.

**Note**: The instructor may choose to inoculate today's culture with the day-old "overnight" stored in the refrigerator.

1. Transfer 10 ml of the sewage-bacteria-bacteriophage culture into a centrifuge tube, and centrifuge the sample at 2000 rpm for 5 minutes. Most of the remaining cells will be pelleted. The supernatant contains bacteriophage.
2. Prepare a 10-ml storage tube for the collection of bacteriophage supernatant as it is filtered. Then pipette the supernatant into a 10 ml syringe barrel fitted with a 0.45-micron filter. Gently slide the plunger, allowing the flow-through to drip into the storage tube. This step removes any remaining bacteria from the phage sample. The storage tube contains bacteriophage. It can be stored at 4°C and is stable for several months.
3. Prepare a series of microfuge tubes for making serial 10-fold dilutions of the bacteriophage suspension (performing the same dilution repeatedly in series is called serial dilution. See Fig. 32.1). Label six tubes 1–6. Into each tube, pipette 0.9 ml of sterile PBS.
4. Perform serial dilutions. Transfer 0.1 ml of phage suspension (that has been mixed well) into tube 1, and mix. Using the same pipette, transfer 0.1 ml of the sample from tube 1 into tube 2, and mix. Repeat this process, transferring 0.1 ml from tube 2 to tube 3, and so on, mixing each time as shown in Fig. 32.1. Store the remaining phage suspension in the refrigerator.
5. Distribute 0.5 ml of log-phase *E. coli* into each of the six microfuge tubes, labelled 1–6.
6. To each tube of bacteria, add 0.1 ml of the corresponding phage dilution (0.1 ml of dilution 6 to cell tube 6, and so on).

**Note**: If you work from the most dilute to the least dilute, you can use the same pipette. Cap the tubes, and mix gently by inverting them.

7. Incubate at 37°C for 10 minutes to allow the phage to adsorb to the bacteria. This is your cell-phage mix.

## OBSERVATION AND RESULT

1. Record the number of plaques on each plate in your laboratory report.
2. Using one of the least – crowded plates, pick an isolated plaque for long-term storage. Pipette 1 ml of PBS into a microfuge tube, and add 1 drop of chloroform. Then, using either the large or small end of a Pasteur pipette (depending on the size of the plaque and the space around it), pierce the agar surrounding the plaque, and pick out the agar "plug" containing the plaque. Place the "plug" agar and all, into 1 ml of PBS. The phage will diffuse into the PBS over time, and the chloroform will kill any remaining bacteria. Store the plaque in the refrigerator (4–10°C).

| Plate number | Plaques per plate | Dilution factor | Volume of phage plated (ml) | Titer calculation= Number of plaques (DF)/ Volume plated (ml) | Titer: plaque forming units per ml |
|---|---|---|---|---|---|
|  |  |  |  |  |  |
|  |  |  |  |  |  |
|  |  |  |  |  |  |
|  |  |  |  |  |  |

8. In the meantime, label six warm, dry, nutrient agar plates 1–6 (one for each infection). Write on the bottom plate along the plate edge. Keep the plates in the 37°C incubator until you are ready to use them.

9. When you are ready to plate cell-phage mixes, collect your warmed, labelled plates from the incubator. Add the contents of cell-phage tube 1 to a vial containing 3 ml of top agarose (molten, at 45°C). Quickly cap the tube, and mix it by gently inverting it three times. Quickly pour the mixture onto the warmed plate 1. You can tilt the plate slightly to spread the top agarose. Push the plate aside, but do not pick it up until the agarose solidifies.

10. Repeat step 9 for each of the remaining five samples 2 to 6.

11. Allow the plates to cool without being disturbed for approximately 10 minutes. When the top agarose has solidified, incubate the plates, inverted, at 37°C for 24 hours.

## REVIEW QUESTIONS

1. Why is bacteriophage titer expressed as PFU/ml and not bacteriophages/ml?

2.  What is the purpose of amplification step?

3.  You picked a single plaque from a phage plate for a long-term storage. It is expected that all of the phages in the storage tube are identical. Why?

4.  Draw and describe how a bacteriophage plaque arises on a bacterial lawn?

# 33

# IDENTIFICATION AND ENUMERATION OF WHITE BLOOD CELLS

## OBJECTIVE

To be able to differentiate between granulocytes and agranulocytes.

## INTRODUCTION

Immunity is the state of protection from an infectious disease, which is achieved by both nonspecific and specific mechanisms. Nonspecific immunity is provided by the skin and mucous membranes, which act as physical barriers to infections, and by several types of white blood cells (leukocytes) some of which are phagocytic, capable of engulfing microorganisms. Specific immunity, on the other hand, is conferred by lymphocytes, including leukocytes that bind specifically to foreign molecules (antigens) on the surfaces of inviting organisms and infected body cells.

There are five major leukocyte types, each classified as either a granulocyte or an agranulocyte. The granulocytes (neutrophils, eosinophils, and basophils) contain cytoplasmic granules that are packed with degradative enzymes and mediators of inflammation. The agranulocytes consist of two morphologically and functionally distinct cell types (monocytes and lymphocytes) that have finer, less prominent granules. A white blood cell can usually be identified by the shape of its nucleus and by the presence or absence of granules. For example, a mature neutrophil has a distinctive multilobed nucleus and fine granules in its cytoplasm, while a lymphocyte has a rounded nucleus that fills much of the cell interior, and no obvious granules.

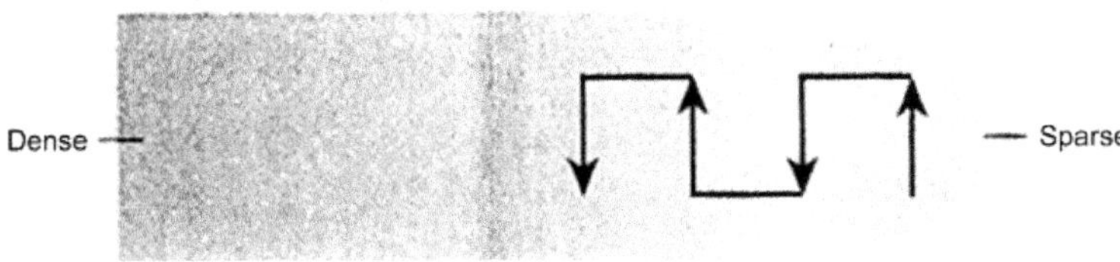


**Figure 33.1**    The cross-section method for counting white blood cells.

If you were to count all of the white blood cells in a particular volume of blood, you would be determining the total white blood cell count, a value expressed in number of cells per micro litre ($\mu$l). If you were to identify each cell type as you count it, you would be generating a differential white blood cell count. A differential count is a measure of each leukocyte type (both mature and immature forms) and is likewise expressed as cells/$\mu$l for each cell type. As you will see, however, the most common cell in a sample is whole blood cell, or erythrocyte. Typical erythrocyte counts range from 45,00,000 to 65,00,000 cells per $\mu$l, while total white blood cell counts range from 4,500 to 11,000 cells per $\mu$l.

The finger-stick method of blood collection provides a small amount of mixed capillary, arteriole, and venule blood. This type of blood collection is frequently used when only a small amount of blood collection is needed. It is also used on infants younger than 6 months of age, in young children, and in adults who have poor veins or whose veins cannot be used because of intravenous infusions. As with any blood collection or invasive procedure, all materials that touch the subject must be sterile (alcohol wipe, cotton ball, lancet, Band-aid), and blood-contaminated materials must be disposed properly. Review the universal precautions, in the laboratory safety section of this manual to learn the steps you must take when handling human source samples.

Although automated methods are now available for identifying and counting white blood cells, you will be identifying and counting cells with the aid of staining and light microscopy.

## MATERIALS

### Blood collection

#### Miscellaneous

 Latex gloves
 Sterile, disposable safety lancets
 Sterile cotton balls
 70% ethanol, or alcohol wipe
 Two clean microscope slides (one slide for sample, one spreader slide)
 Orange biological disposal bag (one per lab)
 Plastic sharps collector (one per lab)

**Wright's staining**

Reagents

> Wright–Giemsa Stain and buffer
> Distilled water
> 70% ethanol

Equipment

> Light microscope

Miscellaneous

> Staining support and drip pan
> Human blood film, smear
> Latex gloves

## PROCEDURE

**Blood collection**

1. Use the middle of the outer segment of the third or fourth finger. To increase local blood flow prior to the puncture, you can wrap the finger in a warm, moist paper towel for 2–3 minutes.
2. Clean the site with an alcohol wipe, and allow it to air-dry. Do not blow on the skin to dry the alcohol. Blowing can contaminate the site. It is important to completely air-dry the residual alcohol because it may cause rapid hemolysis when it contacts the blood.
3. Using the sterile lancet collect your blood sample. Immediately dispose of the cartridge in the plastic sharps collector.
4. Wipe away the first drop of blood which usually contains excess tissue fluid.
5. Place 2 drops of blood near one end of a clean slide. Avoid skin contact with slide. Place the short end of a second, spreader slide into the blood drops, with a 30° to 40° angle between the two slides, until the blood spreads along the edge of the spreader side. Just before the blood has spread completely along the edge of the spreader slide, push the spreader slide along the first slider to from a smear. The smear should show a gradual transition from thick to thin.

6. Label the slide with your name and the date, and allow the smear to air-dry.
7. Dispose of all materials properly.

## Wright's staining

1. Place slides on a staining support over a drip pan, and apply enough drops of Wright–Giemsa stain to cover the smear. Count the number of drops you use. Incubate for 1 minute at room temperature.
2. Add an equal number of drops of Wright's phosphate buffer (pH 6.4). Gently blow on the slide to mix the solutions, and incubate for 3–6 minutes at room temperature.
3. Rinse the slides with distilled water. Cleanse the back of the slide with 70% ethanol.
4. Allow the smear to air-dry.
5. Examine the stained blood smear. Scan the blood smear at 40X magnification. Erythrocytes (red blood cells) are rose to salmon in colour, are biconcave, and have no nucleus. Most of the cells you see are erythrocytes. The leukocytes (white blood cells) are large, nucleated cells, coloured shades of blue by the stain. The nuclei of white cells will be blue to light purple. The cytoplasm will vary from pale pink (neutrophils), to light blue (lymphocytes). Eosinophils and basophils are discernible by their granules (orange to rose for eosinophils; violet to blue for basophils). Neutrophils also contain granules that stain pink to purple.
6. Perform a differential white blood cell count. Shift the magnification to 100X. Identify and record the white blood cell types you see, starting from the sparse end and working toward the denser end of the smear. Use the cross-sectional method of differential counting. Count from the bottom right of the sparse end count up, count left, count down, count left, etc. as shown in Fig. 33.1. When you see a white blood cell, identify and record it in your laboratory report. Design a data table that will allow you to easily record your data. Stop when you have counted a total 100 white blood cells.
7. Calculate the percentage of each leukocyte type you counted. When counting 100 cells, the percentage is easily discerned. A count of 65 neutrophils means that neutrophils make up 65% of the total white cells in the sample.

## RESULTS

| White blood cell type | Labelled diagram of cell type | Number of cells counted | Percent of total cells | Typical percent range for cell type |
|---|---|---|---|---|
|  |  |  |  |  |
|  |  |  |  |  |
|  |  |  |  |  |

## REVIEW QUESTIONS

1.  If you counted 100 total white cells over a portion of a smear that is
    equivalent to about 50 µl of blood, would that white blood cell count
    be within normal range?

# 34

# ANTIGEN–ANTIBODY PRECIPITATION REACTIONS AND DETERMINATION OF ANTIBODY TITER

## OBJECTIVES

1. To become familiar with antigen–antibody precipitation reaction.
2. To screen the causative agent for a specific disease.

## INTRODUCTION

Specific immunity is conferred by T lymphocytes (T cells) and B lymphocytes (B cells), the two special classes of leukocytes that bind to foreign antigens (similar to enzyme–substrate binding) and become activated to help in clearing the antigens from the body. T cell responses include the development of active cytotoxic T cells that destroy virus-infected cells and other abnormal body cells. B cells respond to antigen by developing into plasma cells that secrete high levels of proteins called antibodies or immunoglobulins. Antibodies circulate in the blood and tissue fluid, binding to the precise antigens that induce their production in the first place, and form antibody–antigen complex. Once tethered in a large complex, viruses, bacteria, and toxins are effectively blocked from harming the body and are targeted for destruction by a variety of mechanisms such as phagocytosis.

Antibodies are present in the serum, the viscous yellow fluid that remains after red and white blood cells have been separated from the fluid portion of blood (plasma) and a clot has formed. Because of its antibody content, serum from an infected or immunised person or animal is sometimes called antiserum. Antiserum contains many

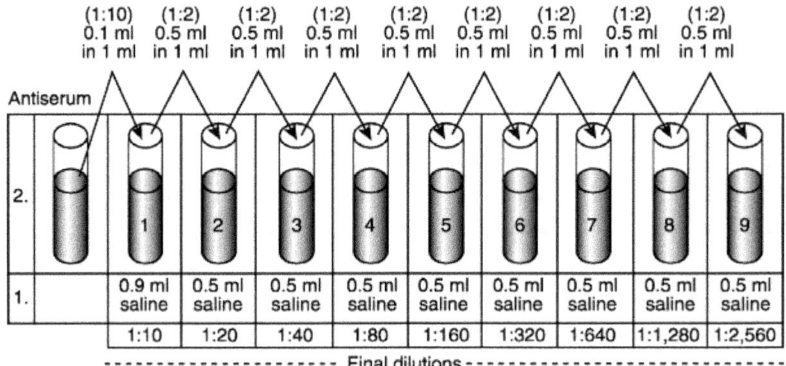


**Figure 34.1**   Serial dilution of antiserum.

antibody molecules specific for the infecting or immunising agent. In fact, the study of antibody–antigen reactions *in vitro* is known as serology.

In serological or immunological methods, antibodies are used as tools for the detection and quantification of specific molecules or microbes. The antibodies collected from the serum of an animal that has been immunised with the selected antigen are called polyclonal antibodies because they are the products of many different B cells (plasma cells) specific for different parts of the same antigen. Monoclonal antibodies (MABs) on the other hand, cannot be collected directly from an animal. MABs are secreted from a cultured hybridoma — a fusion of two cells, one a B cell from an immunised mouse and the other a type of cancer cell called a myeloma. The B cell partner provides the desired antibody specificity, and the myeloma partner contributes immortality to the hybridoma. The hybridoma becomes a continuous source of monoclonal antibodies of singular specificity.

In addition to the appropriate polyclonal or monoclonal antibody preparation, serological methods require a way to detect antibody–antigen binding. Some serological methods take advantage of the formation of visible antibody–antigen complexes (agglutination and precipitation tests). Others make use of antibodies that have been covalently linked to a detectable label (Enzyme-linked immunosorbent assay and Western blotting).

In this exercise, you will use a precipitation test to determine the titer of antibodies in the serum from an immunised animal. An antibody titer is a measure of the relative strength of an antiserum, and is expressed as the reciprocal of the greatest dilution of antiserum still capable of mediating a detectable effect (here, precipitation). For example, if antiserum A activity is detectable up to a dilution of 1 : 200 and antiserum B activity is detectable up to a dilution of 1 : 400, then the strength, or titer, of antiserum B (400) is greater than that of antiserum A (200).

The precipitation test is among the simplest and quickest of the serological methods and is based on the propensity of antibodies to form complexes with their corresponding antigens. When antibodies attach to antigen molecules in solution, the molecules become part of an insoluble antibody–antigen complex, and a visible precipitate forms. Thus, the presence of a precipitate is a positive result, the fact that the antibody molecules have bound specifically to the antigen molecules is obvious. (Similarly, when the antigen is located on a cell surface the cells become clumped together or agglutinated by antibodies). Here, you will use a series of precipitation tests to determine the titer of

## OBSERVATION AND RESULT

Record your results in the laboratory report, indicating the extent of precipitate formation, if any (– (no ppt), +, ++, +++).

Antiserum:

Antigen:

| Dilution | 1 | 2 | 3 | 4 | 5 | 6 | 7 | 8 | 9 |
|---|---|---|---|---|---|---|---|---|---|
| Result |  |  |  |  |  |  |  |  |  |
|  |  |  |  |  |  |  |  |  |  |

a polyclonal antiserum raised in rabbits against the protein albumin from cows (known as BSA, for bovine serum albumin).

## MATERIALS

### Reagents

Antigen: bovine serum albumin (BSA)
(25 µg/ml)
Antiserum: rabbit anti-BSA
Phosphate-buffered saline (PBS)

### Miscellaneous

10 serological (precipitin) tubes (6 × 50 mm)
Precipitin tube rack
1.0 ml serological pipette
Pasteur pipette/bulb
1.5 ml microfuge tubes
Laboratory marker

## PROCEDURE

1. Prepare tubes for antiserum dilutions. Label nine microfuge tubes with a lab marker, numbering them 1 to 9. Using a 1.0 ml serological pipette, transfer 0.9 ml of 0.9% saline into tube 1. Then pipette 0.5 ml of 0.9% saline into each of the remaining tubes, 2 to 9.

2. Perform antiserum dilutions (Fig. 34.1): Pipette 0.1 ml of rabbit anti-BSA into tube 1. Mix by pipetting gently up and down about five times, try to avoid bubbles, and with the same pipette transfer 0.5 ml from tube 1 into tube 2. Again, mix the antiserum and saline in tube 2 by pipetting gently up and down about five times, trying to avoid bubbles. With the same pipette, transfer 0.5 ml from tube 2 to tube 3, and so on, repeating the mixing and transfer of 0.5 ml volumes in series to tube 9. (Tube 9 contains 1.0 ml of diluted antiserum).

3. Label 10 serological tubes, 1–10. Using a Pasteur pipette, transfer 8 drops of 0.9% saline into serological tube 10. Using the same Pasteur pipette, transfer 8 drops of antiserum from dilution tube

9 to serological tube 9. Do the same for samples 8 through 1, in that order (you can use the same Pasteur pipette because you are handling samples from the most dilute one to the least).

4.  The antigen sample (BSA) is provided in 0.9% saline at a concentration of 25 µg/ml. Using a Pasteur pipette, gently layer about 1 ml of the BSA preparation over the antiserum in each of the 10 precipitin tubes. It is important not to mix the two layers together. You are looking for local precipitate forming at the interface, where antigens meet antibodies.

5.  Watch for a reaction in the tubes after you have finished adding the antigen to all of them. Check for the formation of precipitate at the interface every 3 minutes for 15 minutes.

## REVIEW QUESTIONS

1.  What is the titer of the antiserum?

2.  What is the purpose of reaction tube 10?

# 35

# AGGLUTINATION REACTIONS:
# ABO BLOOD TYPING

## OBJECTIVES

1.   To become familiar with
     (a)   Agglutination reaction.
     (b)   Different types of blood groups.
2.   To determine the blood group of an unknown sample.

## INTRODUCTION

The highly specific and selective binding of antibodies to antigens has led to the development of a number of antibody based diagnostic and research methods. In these methods (agglutination, precipitation, ELISA, Western blot, and others), antibodies are used as tools for the detection and quantification of drugs, hormones, and other molecules, as well as for the identification and characterisation of viruses, cells, and tissue. Alternatively, such molecules, viruses, and cells (antigens) can be used to detect the presence of particular antibodies in a test sample. In the HIV screening test, for example, a person's serum is tested for the presence of antibodies specific for HIV, and not for the virus itself. In this case the antigen is the tool or the "known", and the antibody specificity is the "unknown".

When antibodies bind to cells, such as bacteria, yeast, or red blood cells, the cells clump together, or agglutinate. A visible agglutination reaction indicates that antibodies are binding specifically to cells, linking them together to form a large complex. So, just as antibodies bind to soluble molecules to form an insoluble precipitate, they bind to cell-bound molecules to form a clump of cells. Agglutination reactions

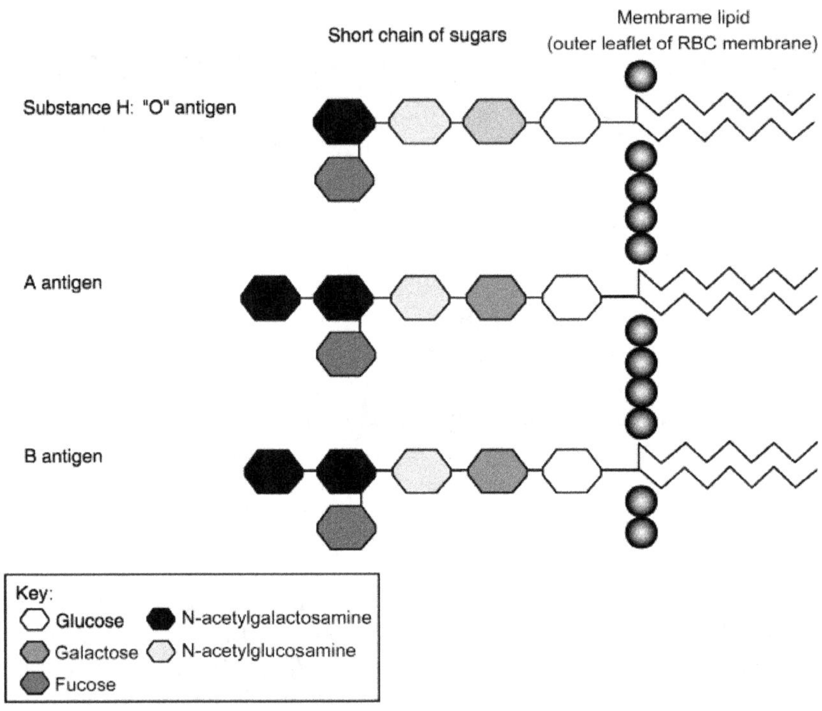


**Figure 35.1**   Illustration of  O, A and B antigens.

are routinely used to type blood, to identify microorganisms, and to test serum samples for the presence of antibodies reactive against a particular microbe.

In blood typing, antibodies are used to detect red blood cell surface antigens such as those of the ABO system. These antigens consist of a core glycolipid called substance H. If substance H is modified by the addition of another sugar, N-acetylgalactosamine, it is an *A* antigen. If substance H instead has an attached galactose, it is a *B* antigen. If substance H stands alone, unmodified, it is neither *A* nor *B*, and is known as *O* (Fig. 35.1). *A* antigens are found on type *A* and type *AB* red blood cells, and neither is found in type *O* red blood cells. Each of these blood types may be $Rh^+$ or $Rh^-$, depending on the presence or absence of another antigen, a protein called the Rh factor.

It may seem odd that normal molecules such as these are called antigens. In fact, all macromolecules are potential antigens, especially if they are transferred into a nonidentical person or to another animal through transfusion or transplantation. In autoimmune disorders, normal self-molecules such as these may be treated as foreign antigens. So macromolecules can be called antigens because they have the potential to generate antibodies.

In this exercise, you will determine the blood type of either an aseptic blood sample or your own blood sample using the agglutination reaction.


## MATERIALS

### Blood collection

#### Miscellaneous

    Latex gloves
    Sterile, disposable safety lancets
    Sterile cotton balls
    70% ethanol, or alcohol wipe
    Two clean microscope slides (one slide for sample, one spreader slide)
    Orange biological disposal bag (one per lab)
    Plastic sharps collector (one per lab)

## OBSERVATION AND RESULTS

Record your agglutination results

| Sample | Anti-A reaction | Anti-B reaction | Anti-Rh reaction | Phenotype (blood type) | Genotype |
|---|---|---|---|---|---|
|  |  |  |  |  |  |
|  |  |  |  |  |  |
|  |  |  |  |  |  |
|  |  |  |  |  |  |

### Aseptic blood and ABO – Rh blood typing

Reagents

Aseptic blood samples, anti-A, B, and Rh
Antisera and materials provided in a blood cell/antisera biokit

Miscellaneous supplies

Latex gloves

## PROCEDURE

### Blood collection

1.  Have a test card ready. Label it with your name and the date. If you are using an aseptic blood sample, go to step 5. To prepare for the blood collection consider using the middle of the outer segment of the third or fourth finger. To increase local blood flow prior to the puncture, you can wrap the finger in a warm, moist paper towel for 2–3 minutes.
2.  Clean the site with an alcohol wipe, and allow it to air-dry. Do not blow on the skin to dry the alcohol. Blowing can contaminate the site. It is important to completely air-dry the residual alcohol because it may cause rapid haemolysis when it comes in contact with the blood.
3.  Using the sterile lancet needle, collect the blood sample. Immediately dispose of the cartridge in the plastic sharps collector.
4.  Wipe away the first drop of blood with the sterile cotton ball. The first drop of blood usually contains excess tissue fluid. Allow the blood to drop onto each of the three areas on the test card marked "blood". Go to step 6.

### ABO-Rh blood typing

5.  If you are using a provided, aseptic blood sample, place 1 drop of the blood on each of the three areas on the test card marked "blood".
6.  Place one drop each of the anti-A serum, anti- B serum, and anti-D (Rh) serum onto the designated areas of the test card.
7.  Mix each of the blood drop–antiserum sets, using a fresh mixing stick for each sample. Dispose of each mixing stock in the sharps container as soon as you are done with it.

8.  Gently rock the test card for 1 minute, without allowing the samples to flow out of their designated test areas.

9.  Examine the samples for agglutination. The anti-D reaction may take a longer time than the other. It may be easier to see agglutination if you tilt the card slightly.


## REVIEW  QUESTIONS

1.  What are the different types of blood grouping systems available?

2.  What is Bombay blood group?

3.  What do you mean by Rh factor?

# 36

# IMMUNODIFFUSION: ANTIGEN–ANTIBODY PRECIPITATION REACTIONS IN GELS

## OBJECTIVE

To become familiar with various immunodiffusion techniques.

### INTRODUCTION

When soluble antigen molecules become linked together by multiple antibodies, an insoluble precipitate forms. This precipitate, visible to the naked eye, can reveal the identity of an antigen (if the antibody specificity is known) or the specificity of the antibody (if the antigen is known). A precipitate also indicates that the antibody and antigen molecules are present at optimal proportions for the formation of a large complex, or lattice. In this equivalence zone, there are about two to three antibody molecules for every antigen molecule, leaving no free antigens or antibodies.

In immunodiffusion tests, antibodies and/or soluble antigens are loaded into separate wells of a gel and are allowed to diffuse, each reagent moving radially into the gel. An immobile precipitate, visible as a band (precipitation line) in the gel, develops if specific antibody–antigen binding takes place, and if antibody and antigen components are present at optimal proportions. Double immunodiffusion, also known as Ouchterlony, is the most widely used gel precipitation technique in the research laboratory, while radial immunodiffusion and immunoelectrophoresis are principally used in clinical labs to test serum protein levels.

In double immunodiffusion, antigen and antibody preparations are loaded into separate wells of an agarose gel. In this example,

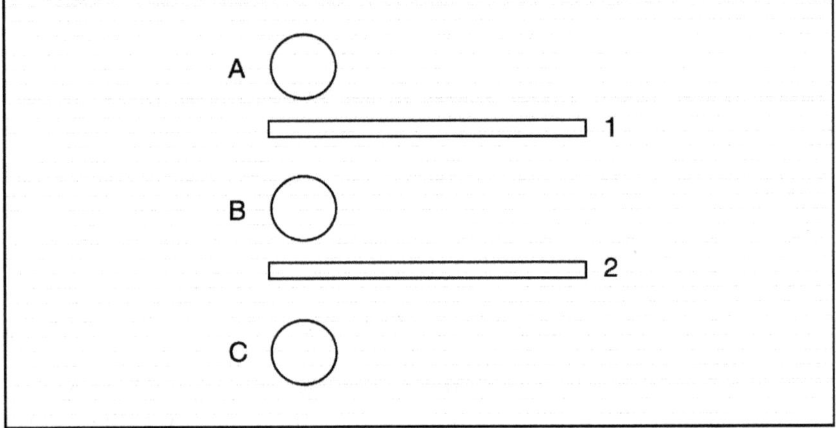

**Figure 36.1**   IEP Assay.

the antibodies (specific for human serum proteins) are located in the centre well, and the antigens (serum proteins) are located in the outer wells. Each substance diffuses from its well, and in time, white lines of insoluble precipitate appear at positions where antibodies have bound to their specific antigens at optimal proportions (the equivalence zone). In radial immunodiffusion, the antibodies are evenly distributed within a preformed 1% agarose gel. Therefore, in this test only the antigen sample loaded into a well diffuses. As antigen molecules move through the gel, they bind to and carry the antibodies, until the ratio of antigens to antibodies is optimal for complex formation. At this point, a ring of precipitation forms, and its diameter is proportional to the concentration of antigen loaded into the well. At higher concentrations, the diameter of the ring is greater because antigen molecules must migrate farther before they gather up enough antibodies to form a complex.

The third type of immunodiffusion is called immunoelectrophoresis. Antigens are first loaded into wells of an agarose gel and are separated by charge in an electrophoresis chamber. Antibodies are then used to detect the separated antigens; after being loaded into a trough that runs the length of the gel, they diffuse toward and complex with the antigens, and form visible lines of precipitate.

## MATERIALS

### Double immunodiffusion (Ouchterlony)

**Note**: As an alternative to the following reagents and procedure, the Ouchterlony procedure can be accomplished using a kit.

### Reagents

Agarose: 40 ml 1% (w/v) molten agarose in 0.05 M Tris-C1, pH 8.6

### Antibodies (Table 36.1)

Serum antibody set
Goat anti-bovine albumin
Goat anti-horse albumin
Goat anti-swine albumin

## Antigens

Serum antigen set — bovine serum, horse serum, swine serum

## Equipment

Microwave oven
Water bath at 55°C

## Miscellaneous

60-mm diameter petri dishes
Covered box for gel storage
Label tape
Laboratory marker
10-ml pipette
Glass dropper (well cutter)
Micropipette/tips (1–10 µl)

## Radial immunodiffusion

Reagents
Human IgG, IgA, and IgM
Radial immunodiffusion kit
Human serum

## Miscellaneous

Micropipette/tips (1 – 10 µl)

**Table 36.1**  Sample loading Order for Double Immunodiffusion Assays

|  | Centre well* | Outer wells* |
|---|---|---|
| Pattern A | Goat anti-bovine albumin | 1. Bovine serum |
| Pattern B | Goat anti-horse albumin | 2. Horse serum |
| Pattern C | Goat anti-swine albumin and goat anti-bovine albumin | 3. Swine serum |
|  |  | 4. Swine serum |

*The contents of the outer wells are the same for all three assays.
**The centre well antibodies are different for each assay.

## Immunoelectrophoresis

### Reagents

High-resolution electrophoresis buffer, pH 8.8
1% agarose in high-resolution buffer, pH 8.8
Antigens: bovine serum, bovine albumin, 10 mg/ml
Antibodies: anti-bovine albumin, anti-bovine serum

### Equipment

Horizontal gel electrophoresis box
Power supply

### Miscellaneous

60 mm diameter petri dish
Tape
2 glass slides
Glass or plastic dropper (well cutter)
Micropipette/tips (10–100 µl)
Grade No. 1 Whatman paper

## PROCEDURE

### Double immunodiffusion (Ouchterlony)

1. Prepare 40 ml of 1% agarose. Add 0.4 g of agarose to 40 ml of 0.05 M Tris-C1, pH 8.6 in a 125-ml flask. Boil the mixture for about 30 seconds, checking to make sure that it does not boil over. Using a hot glove, gently swirl the flask, and return it to the oven. Heat for 15 seconds, repeating this until no flecks of agarose are visible in the flask. Let the molten agarose cool until the flask is comfortable to handle, but still warm.

2. Obtain three 60-mm diameter petri dishes. Writing with a lab marker on the plate bottom, label the three plates A, B and C respectively. Write your initials on all three plates. Pipette 5 ml of slightly cooled molten agarose into each dish. Allow the agarose to solidify, for about 20 minutes.

3. Using the large end of a plastic or glass dropper (a diameter of about 0.5 cm), cut wells into each gel as shown in the following template.

4.  Label the outer wells 1,2,3, and 4 by writing on the plate bottom.
5.  Changing micropipette tips between different reagents, pipette 20 µl of the appropriate antigen and 20 µl of antibody to the designated wells according to the loading order in Table 36.1.
6.  Line the bottom of the storage box with a moist paper towel, and place the dishes in the storage container. Make sure the dishes are level. Incubate the gels for 24–48 hours at room temperature to allow diffusion and banding. The gels can be stored in the refrigerator for several weeks if the box is kept moist.

## Radial immunodiffusion

1.  Using a micropipette, obtain 5 µl of human serum, and pipette it into a designated well of the RID assay gel.
2.  Reserve three wells for standard concentrations, namely a high standard, a low standard, and a serum control (provided in the RID kit).
3.  Place the gel into the moist box, and incubate for 24 hours at room temperature to allow diffusion and banding. Again, the gels can be stored in the refrigerator for several weeks if the box is kept moist.
4.  Read the results of your assay. Measure the diameter of the circle of precipitate (in centimetres) for the sample you have loaded and for the three standard samples. Record these results in your laboratory report.

## Immunoelecrrophoresis

1.  Prepare 40 ml of 1% agarose. Add 0.4 of agarose to 40 ml of high-resolution buffer, pH 8.8, in a 125-ml flask. Microwave the mixture for about 30 seconds, checking to make sure it does not boil over. Using a hot glove, gently swirl the flask, and return it to the microwave. Heat for 15 seconds, repeating this until no flecks of agarose are visible in the flask. Let the molten agarose cool until the flask is comfortable to handle, but still warm.
2.  While the agarose cools, prepare a horizontal gel electrophoresis box by putting the dams securely in place. Also prepare a trough-forming apparatus. Obtain a 60-mm petri dish with lid, and tape a slide to each side.
3.  Once the agarose has cooled so that the flask is comfortable to hold, pour the agarose into the unit until it completely covers the platform. Place the trough-forming apparatus at the center of the platform. Allow the agarose to solidify, for about 10 minutes.

4.   When the gel is solid, gently remove the trough-forming apparatus. Using the large end of a plastic or glass dropper (a diameter of about 0.5 cm), cut wells into each gel as shown in the following template, (Fig. 36.1).

5.   Pour high-resolution buffer, pH 8.8 into the electrophoresis box on either side of the gel, being careful not to pour onto the gel itself, 1 or 2 inches deep. Cut two pieces of Whatman chromatography paper wicks, and place them into the apparatus. Be sure that the paper is in contact with the gel and the buffer at both ends of the gel.

6.   Load the antigens. Changing the micropipette tips between samples, load 20 ml of bovine serum into wells A and C, and 20 ml of bovine albumin into well B. Do not load the troughs.

7.   Electrophorese samples at 70 volts for 1.5 hours.

8.   Load the antibodies. After electrophoresis is complete, load 50 μl of anti-bovine albumin into trough 1 and 50 μl of anti-bovine serum into trough 2. Again, change tips between samples.

9.   Leave the gel in the electrophoresis apparatus, and wrap a moist paper towel and plastic wrap around it to create a moist container. Incubate for 24–48 hours at room temperature to allow diffusion and banding.

## OBSERVATION AND RESULT

### Double immunodiffusion (Ouchterlony)

1.   A precipitin line represents a specific antibody–antigen reaction occurring between the antibodies diffusing from one of the outer wells. The precipitin line should be perpendicular to an imaginary straight line drawn from an outer well to the centre well. The predicted results for pattern A are shown here. Predict the results for patterns B and C.

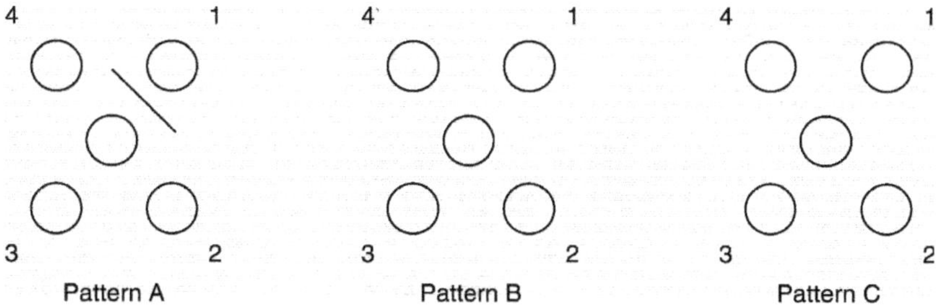

2.  The double immunodiffusion assay can also be done to test the relatedness of antigens loaded into adjacent wells. If the two antigen samples are identical, a smooth corner forms where the two lines meet (identity). If the two antigen samples are not identical but related, then a spur forms at the corner (partial identity). Finally, if the two adjacent antigen samples are not related at all, two spurs form at the corner (nonidentity).

Antibodies specific for all human serum proteins were loaded in the centre well. No reaction is expected between the centre well and the antigen wells containing hen egg lysozyme or cytochrome *c*.

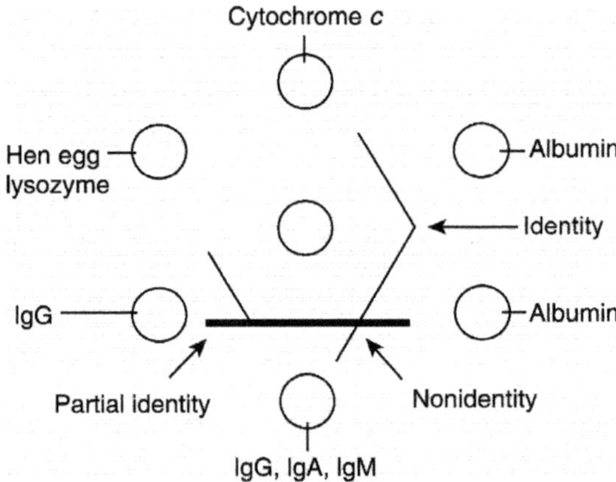


3.  Return to question 1, and consider whether spurs should be included in your predicted results. Add spurs to the diagram if they are expected.
4.  Draw your double immunodiffusion results. Do they agree with your predictions?

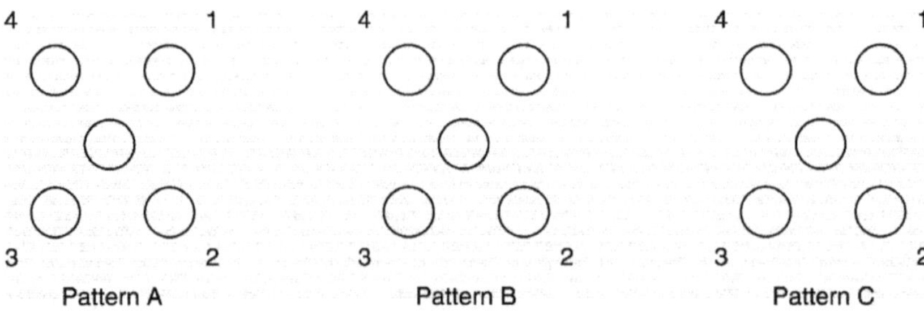

### Radial immunodiffusion

1.  Record the three standard immunoglobulin (Ig) concentrations and the diameter of each resulting precipitin ring. Also record the diameter of the precipitin circle for the sample you have loaded.

| Ig standard concentration | Diameter of precipitate circle |
|---|---|
|  |  |
|  |  |
|  |  |

2.  *Graph the standard results.* On a semilog paper, plot the three known, standard Ig concentrations (on the log scale) versus the diameter of the corresponding precipitin rings (on the linear scale). If you did not run standards, analyse the gel shown in Fig. 43.3*a*.

Comment on your results. The normal mean concentration of each antibody in serum is presented in Table 36.2. The radial immunodiffusion test is often done to determine the concentrations of IgG, IgA, and IgM.

**Table 36.2**   The five classes or *isotypes* of antibodies (immunoglobulins)

| Antibody isotype | Mean concentration (mg/ml) |
|---|---|
| IgG | 13.5 |
| IgA | 3.5 |
| IgM | 1.5 |
| IgD | 0.03 |
| IgE | 0.0005 |

### Immunoelectrophoresis

1.  Draw the results of the IEP assay.

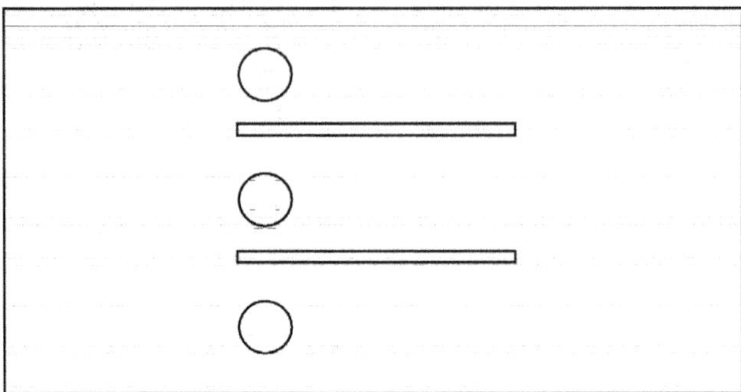

2.  The following diagrams represent the results of two IEP assays done on the serum of a young child who has experienced frequent infections since infancy. What is your diagnosis?

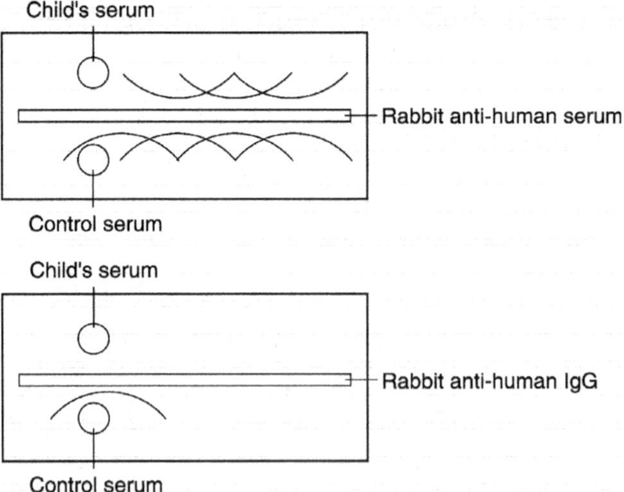

# 37

# ENZYME-LINKED IMMUNOSORBENT ASSAY (ELISA)

## OBJECTIVE

To demonstrate a method for the identification of either an antigen or an antibody by use of an enzyme-labelled antibody test procedure.

## INTRODUCTION

In this laboratory experiment, we will simulate the transmission of a hypothetical infectious disease among members of the class. The infectious agent or antigen we are using is a harmless protein. However, for our purposes, consider it contagious and dangerous. The transmission of this hypothetical disease sets the stage for an ELISA (enzyme-linked immunosorbent assay), an antibody-based test that is commonly used to study the specific attachment that occurs between an antibody and an antigen (thus the term immunosorbent). It is enzyme-linked because an enzyme is covalently attached to the tail portion of the antibody. The enzyme linked to the antibody is one that catalyses the conversion of a colourless substrate into a coloured product.

In an ELISA, the test sample, here simulated body fluid is loaded into one well of a 96-well micro titer plate. Next, the enzyme-linked antibody specific for the infectious agent is added to each well. After washing to remove nonspecifically bound antibodies, the chromogenic (colour-generating) substrate is added. The development of colour in a well indicates a positive result, whereas if the sample remains colourless, it is negative. In addition, the intensity of colour is an indication of the amount of reaction product in that well, which in turn correlates with the amount of enzyme-linked antibody and so, the concentration of antigen in the well.

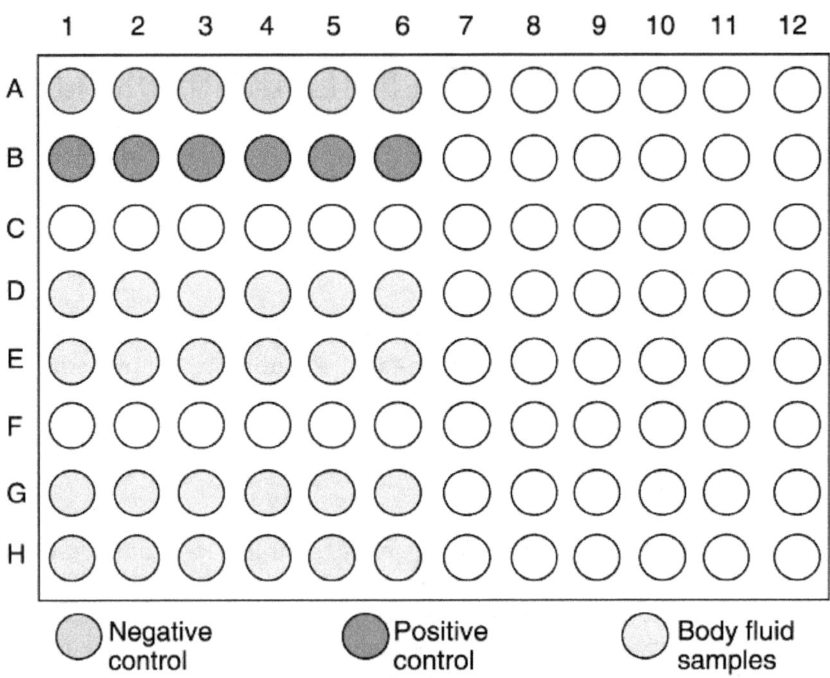


**Figure 37.1**  Sample loading on an Elisa plate.

| To wells | Add 2 drops or 100 $\mu$l of |
| --- | --- |
| A1–A6 (row A) | negative control |
| B1–B6 (row B) | positive control |
| D1–D6 (row D) | nonsharing fluid (partner A) |
| E1–E6 (row E) | sharing fluid (partner A) |
| G1–G6 (row G) | nonsharing fluid (partner B) |
| H1–H6 (row H) | sharing fluid (partner B) |

Like the precipitin reactions (ID, IEP, and RID) and agglutination reactions, the ELISA takes advantages of a specific interaction between an antibody and an antigen, but unlike these, detection by ELISA doesn't require the formation of large antibody–antigen complex. Therefore, the ELISA is much more sensitive than precipitin-type tests.

## MATERIALS

### Exchange of "body fluid"

#### Reagents (see Table 37.1)

Two microfuge tubes that contain the solution representing body fluid (per person)
1.0 ml serological pipette or Pasteur Pipette/bulb

### Analysis of samples by ELISA

#### Reagents (per pair)

Positive control (contains infectious agent)
Negative control (contains no infectious agent)
Nonsharing fluid (partner A sample)
Sharing fluid (partner A sample)
Nonsharing fluid (partner B sample)
Sharing fluid (partner B sample)
Washing buffer
Enzyme-linked antibody reagent
Substrate (colour-change reagent)

#### Miscellaneous

96-well ELISA microtiter plate
Micropipette/tips (100 µl)
10-ml pipette
Pasteur pipettes/bulb
One piece of bench-coat absorbent paper

**Table 37.1**   Table of Reagent Recipes for Simulated infectious disease transmission

| Simulated substance | Identify | Recipe |
| --- | --- | --- |
| Body fluid | Sodium carbonate buffer | 0.16 g sodium carbonate |
| | | 0.27 g sodium bicarbonate in 100 ml distilled water |
| Viral antigen | Biotinylated albumin | 10 l of biotinylated albumin at 6 mg/ml in 10 ml of sodium carbonate buffer |
| Wash buffer | PBS 0.1% Tween-20 | 32 g sodium chloride |
| | | 0.8 g potassium chloride |
| | | 4.48 g sodium phosphate, dibasic |
| | | 0.8 g potassium phosphate, monobasic |
| | | 2 ml Tween-20 |
| | | Distilled water to 2 litres |
| Enzyme-linked Antiviral antibody | Streptavidin peroxidase (Sigma-Aldrich #5512) | 5 l of streptavidin peroxidase (0.5 mg/ml 50% glycerol) in 50 ml of wash buffer |
| Substrate | TMB (tetramethylbenzidine) in phosphate citrate solution | Phosphate citrate solution: Combine 25.7 ml |
| | | 0.2 M dibasic sodium phosphate and 24.3 ml 0.1 M citric acid solution with 50 ml distilled water |
| | TMB tablets (Sigma-Aldrich #T 3405) | Dissolve 3 mg TMB in 30 ml of phosphate citrate solution. Add 5 l 30% hydrogen peroxide. (Use same day: keep cold and dark) |

## PROCEDURE

### Exchange of "body fluids"

1.   Label the two body fluid tubes with your name, and place one of them (labelled "nonsharing") in the rack at the front of the room. Use the second body fluid tube (labelled "sharing") for the following steps.

Proceed with steps 2 through 4 using your "sharing" tube. At the end of each exchange, you should have about the same volume of fluid you started with.

**Table 37.2**  ELISA Results and Potential Transmission Events

| Name | Sharing ELISA results | Exchange 1 | Exchange 2 | Exchange 3 | Analysis (excluded or not excluded as original carrier) | Non sharing ELISA results |
|---|---|---|---|---|---|---|
|  |  |  |  |  |  |  |
|  |  |  |  |  |  |  |
|  |  |  |  |  |  |  |
|  |  |  |  |  |  |  |
|  |  |  |  |  |  |  |
|  |  |  |  |  |  |  |
|  |  |  |  |  |  |  |
|  |  |  |  |  |  |  |
|  |  |  |  |  |  |  |
|  |  |  |  |  |  |  |
|  |  |  |  |  |  |  |
|  |  |  |  |  |  |  |
|  |  |  |  |  |  |  |
|  |  |  |  |  |  |  |
|  |  |  |  |  |  |  |
|  |  |  |  |  |  |  |
|  |  |  |  |  |  |  |
|  |  |  |  |  |  |  |
|  |  |  |  |  |  |  |
|  |  |  |  |  |  |  |
|  |  |  |  |  |  |  |
|  |  |  |  |  |  |  |
|  |  |  |  |  |  |  |
|  |  |  |  |  |  |  |
|  |  |  |  |  |  |  |
|  |  |  |  |  |  |  |
|  |  |  |  |  |  |  |
|  |  |  |  |  |  |  |
|  |  |  |  |  |  |  |

## OBSERVATION AND RESULT

1. Record your results, providing your name, the results of the ELISA of your own "sharing sample" (+ or −), and the three people with whom you exchanged fluid, in order.
2. Join with two other pairs of students, as a group, work through the path of transmission to determine who the original carrier(s) might be.
3. Working in your original pairs, go back to your plate and add 2 drops (or 100 m1) of substrate to the wells in rows D and G (D1–6 and G1–6). Examine the sample results as in step 12. Remember that these are the original "nonsharing" fluid samples.
4. Record the results of the nonsharing fluid samples. Determine if your group's conclusions were correct regarding the original carrier(s).

2. Using a transfer pipette, exchange about one-half of your sharing fluid with another person in the room. In Table 37.2, record the name of the person with whom you first made contact.
3. At the instructor's signal, find a different person to exchange one-half of your sharing fluid. Record the name of your second contact.
4. At the instructor's signal, find a third person to exchange about one-half of your fluid. Record the name of your third contact.

**Analysis of samples by ELISA**

1. Join with a partner to proceed with the ELISA. The ELISA is designed to establish whether or not the infectious agent is present in your body fluid samples.
2. Write your names or initials on the plate edge. If you are using a Pasteur pipette to add samples and reagents to the ELISA wells, pipette 2 drops of each sample or reagent. Always change pipettes between different samples and reagents. If you are using a micropipette to add samples and reagents, set it at 100 µl. Always change tips between different samples and reagents. Load controls and "body fluid" samples as shown in the list below Fig. 37.1. Pipette carefully and accurately.
3. Incubate the samples at room temperature for 10 minutes, undisturbed.
4. Discard the liquid contents into the sink, and then place the plate facedown on absorbent paper with some force to remove any remaining liquid from the wells.
5. Using a 10-ml pipette, fill each of the wells that you used with wash buffer. Discard the wash solution into the sink.
6. Repeat step 5 twice (for a total of three washes).
7. Place the plate face down on absorbent paper with some force to remove any remaining liquid from the wells.
8. Add 2 drops (or 100 µl) of enzyme-linked antibody to each of the wells.
9. Incubate the samples at room temperature for 10 minutes, undisturbed.
10. Wash the plate thoroughly by repeating steps 4–7.
11. Add 2 drops (or 100 ml) of substrate to each of the wells you used except those in rows D and G. Rows D and G contain nonsharing fluids and will be assayed later.
12. After about 5 minutes of development, examine the qualitative results with respect to sample colour changes.

## REVIEW QUESTION

1.  In the ELISA, what would happen if you eliminate the washing step
    prior to adding substrate?

# 38

# SDS – POLYACRYLAMIDE GEL ELECTROPHORESIS

## OBJECTIVE

To become familiar with the preparation of polyacrylamide gels.

## INTRODUCTION

One-dimensional SDS - PAGE is most often used to denature proteins and separate them based on their size. Since proteins are positively or negatively charged, SDS, an anionic detergent, is included which coats all the proteins giving them a net negative charge. The proteins will travel in the polyacrylamide towards the positive electrode during electrophoresis. A second component of the sample buffer is dithiothreitol (DTT) that reduces the disulphide bonds to assist with the denaturation of the proteins into their subunits. The sample buffer also contains glycerol, which increases the density of the sample for facilitating the loading of samples into the well and bromophenol blue which assists in monitoring the progress of electrophoresis.

## MATERIALS:

**Apparatus** *Vertical slab gel system, mini model.*

**Electrophoresis power supply** *Analytical model.*

***Acrylamide monomer*** *(Stock solution 30 : 0.8)*

Acrylamide                                    30 g
Bisacrylamide                                 0.8 g

   Add distilled water to make the final volume to 100 ml. Store it in a brown bottle.

***Separating gel buffer*** *(4X)  pH 8.8*

Tris                                          18.15 g
Distilled water                               75 ml

   Adjust the pH with 0.1 M HCl and make the final volume to 100 ml with distilled water.

***Stacking gel buffer*** *(4X) pH 6.8*

Tris                                          1.5 g
Distilled water                               20 ml

   Adjust the pH with 1 M HCl and make the final volume to 25 ml with distilled water.

*SDS 10%*

SDS                                           1 g
Distilled water                               10 ml

***Ammonium per sulphate*** *(APS)  10% freshly prepared*

APS                                           0.1 g
Distilled water                               1 ml

***Gel overlay solution*** *pH 8.8*

Separating gel buffer                         25 ml
10% SDS solution                              1 ml
Distilled water                               74 ml

   Make the final volume to 100 ml

*Sample buffer (2X) pH 6.8*

| | |
|---|---|
| Stacking gel buffer | 2.5 ml |
| 10% SDS solution | 4 ml |
| 20% Glycerol | 2 ml |
| 0.02% Bromophenol blue | 2 mg |
| 0.2 M Dithiothreitol | 0.31 mg |

Add distilled water to make the final volume to 1l. Store at room temperature for up to 1 month.

*Water saturated n-Butanol*

| | |
|---|---|
| n-Butanol | 100 ml |
| Distilled water | 10 ml |

Mix them in a bottle and shake well. Use the top layer for overlaying gels. Store at room temperature indefinitely.

*Staining solution*

| | |
|---|---|
| Coomassie Brilliant Blue | 100 mg |
| Methanol | 40 ml |
| Glacial acetic acid | 10 ml |
| Distilled water | 50 ml |

*Destaining solution*

| | |
|---|---|
| Methanol | 40 ml |
| Glacial acetic acid | 10 ml |
| Distilled water | 50 ml |

## PROCEDURE

1. Place a thoroughly cleaned rectangular plate on a table and keep three gaskets.
2. Place the notched glass plate over the gaskets and insert two spacers along the gaskets.
3. Using metal clips, clamp together the rectangular and notched glass plate assembly.

RESULTS

Determine the molecular weight of the unknown protein sample from the standard graph by calculating the Rf value.

4. Keep the whole assembly in a vertical position by placing them in a gel casting stand. Tighten it gently.

5. Prepare the stacking gel mixture (10%)

| | |
|---|---|
| Acrylamide stock solution | 10 ml |
| 4X separating gel buffer | 7.5 ml |
| 10% SDS solution | 0.3 ml |
| Distilled water | 12.1 ml |
| 10% APS | 150 µl |
| TEMED | 10 µl |

6. Using a 10-ml pipette, pour the gel mixture immediately into the glass sandwich up to the mark and allow no air bubbles.

7. Immediately after pouring the separating gel mixture, take about 0.5–1.0 ml of n-butanol saturated with water or separating gel overlay solution and inject it slowly in such a way that the overlay solution spreads over the entire top surface of the separating gel mixture.

8. Allow the gel undisturbed for polymerisation for about 15–20 minutes.

9. After polymerisation tilt the sandwich and discard the overlay solution.

10. Prepare the stacking gel mixture

| | |
|---|---|
| Acrylamide stock | 1.33 ml |
| 4X stacking gel buffer | 2.5 ml |
| 10% SDS solution | 0.1 ml |
| Distilled water | 6.0 ml |
| 10% APS | 50 µl |
| TEMED | 5 µl |

11. Insert the comb and pour the stacking gel mixture. Tilt the assembly for air bubbles (if any) sticking onto the comb to escape.

12. Allow the polymerisation to complete for 15–30 minutes.

13. *Sample preparation* Mix equal volume of protein samples (5 µl) with the 2X sample buffer and keep in a boiling water-bath for 90 seconds. Place the sample in ice till loading.

14. Carefully remove the comb, rinse the wells with the tank buffer and remove the bottom spacer.

15. Clamp the gel plate and tighten it.

16. Using a Hamilton syringe or micropipette apply 10 µl of the sample to each well.

17. Slowly fill the top chamber with tank buffer and add equal volume of tank buffer to the bottom reservoir.

18. Connect the tank to the power supply and turn the power on. Apply 10–15 mA current.
19. Continue the run till the marker dye reaches the bottom of the gel.
20. Turn off the power and disconnect the power cord.
21. Drain the upper reservoir buffer and take the gel out carefully, remove the top notched plate.
22. Place the gel in staining solution, pour enough stain to cover the gel and stain it for 3–4 hrs.
23. Remove the staining solution and add the destaining solution. Shake it intermittently and change the destaining solution several times until a clear background is observed.
24. Examine the bands in a white light transilluminator.
25. Using a ruler measure the distance travelled by the marker dye and each protein from the origin.
26. Calculate the Relative mobility using the formula

$$Rf = \frac{\text{Distance travelled by the protein}}{\text{Distance travelled by the marker dye}}.$$

27. Plot on a semilog graph, MW of standard proteins along y-axis and their Rf values along x-axis.


REVIEW QUESTIONS

1. What is the use of

   a. TEMED



   b. β-mercaptoethanol



   c. Ammonium persulphate

# 39

# WESTERN BLOTTING

## OBJECTIVES

1. To become familiar with the preparation of polyacrylamide gels.
2. To transfer the proteins to nitrocellulose membrane.

## INTRODUCTION

Western blotting involves transfer and immobilisation of proteins from a completed polyacrylamide gel into a nitrocellulose membrane. It is based on the principles similar to Southern blotting. Buffer is drawn through the polyacrylamide gel to the nitrocellulose membrane by capillary action. The proteins are transferred to the membrane in a pattern mirroring their resolution in the gel. Western blot analysis is coupled with immunodetection i.e. using a primary antibody that binds specifically to the protein of interest, allowing that protein to be picked out from all other proteins synthesised by a cell. For the binding of antigen to an antibody to be visible, a chemical tag is required. The tag is a reagent that releases light, catalyses a chemical reaction releasing light or radioactivity. If there is no antigen–antibody reaction no light or radioactivity will be emitted.

Sometimes this tag is conjugated directly to the primary antibody. Alternatively the tag may be bound to a second antibody, one specific for the animal species in which the primary antibody was produced called the secondary antibody. The secondary antibody is used in detection to show that the primary antibody has bound to its antigen. The commonly used tag is alkaline phosphatase and the substrate for this enzyme is 5-bromo-4-chloro-3-indolylphosphate (BCIP).

## MATERIALS

Multiblot electrophoretic transfer unit
Whatman filter papers
Nitrocellulose paper

Transfer buffer (pH 8.3–8.4)

| Tris base | 18.2 g |
|-----------|--------|
| Glycine | 86.5 g |
| Methanol | 1200 ml |
| Distilled water | 4 litres |

Make the volume up to 6 litres.

## METHOD

1. Separate the sample by electrophoresing in 12% polyacrylamide gels as described in the previous experiment.
2. Equilibrate the gel after electrophoresis with transfer buffer for 5 minutes in a trough. The filter papers should also be equilibrated in the transfer buffer for some time.
3. Cut a piece of nitrocellulose paper to the size of the gel and immerse in the buffer.
4. On one lid of the transfer cassette, first place three bits of filter paper and then place the gel carefully on the filter paper.
5. Place the nitrocellulose paper carefully on the gel without entrapping air bubbles between the gel and the nitrocellulose paper.
6. Place another set of filter papers on the nitrocellulose paper.
7. Close the cassette with another lid and place in the transfer tank filled with transfer buffer.
8. Electroblot the sandwich with the gel side facing the cathode and nitrocellulose paper side facing the anode.
9. Carry out the blotting overnight at 20 mA in a cold room.

### Immunodetection

Nitrocellulose papers with the transferred proteins are allowed to react with diluted antisera. Then the antigen–antibody complexes bound to the nitrocellulose paper are allowed to react with alkaline phosphatase conjugated antirabbit – IgG. The colour is then developed by reaction with BCIP substrate to identify the presence of antigen in the sample.

## MATERIALS

Tris buffered saline (TBS)

| | |
|---|---|
| Tris.Cl (pH 7.5) | 100 mM |
| NaCl | 0.9% (w/v) |

Tris buffered saline with Tween 20 (0.02%) - TBST

Blocking buffer
   2% BSA in TBS containing sodium azide (0.02%)

Primary antibody
   1:1000 dilution of antiserum in fresh blocking solution
   1:1000 dilution of pre immune serum

Secondary antibody
   1:25,000 dilution of alkaline phosphatase conjugated antirabbit IgG in blocking solution.

AP buffer (pH 9.5)

| | |
|---|---|
| Tris.Cl | 100 mM |
| NaCl | 100 mM |
| $MgCl_2$ | 5 mM |

AP substrate
   Readymade commercial substrate

| | |
|---|---|
|    EDTA | 20 mM in TBS |

## METHOD

1. Wash nitrocellulose paper in TBS and incubate it in blocking solution for 1 hour at room temperature. After the incubation period wash thrice in TBST for 1 hour with constant shaking.
2. Incubate the nitrocellulose paper with experimental antiserum at 37°C for 1–2 hours with constant shaking and with wash buffer for 1 hour with four changes.

## OBSERVATION AND RESULT

Observe which lane is positive in the western blot.

3. Incubate the nitrocellulose paper with alkaline phosphatase conjugated secondary antibody for 1 hour with constant shaking at 37°C and wash with wash buffer for 1 hour with four changes.
4. Equilibrate the nitrocellulose paper in AP buffer for 10 minutes and damp dry on the Whatman filter paper.
5. Transfer nitrocellulose paper immediately into 10 ml of BCIP substrate solution.
6. Observe for coloured bands in the reactive areas within 10–15 minutes.
7. Wash the nitrocellulose paper thoroughly with water and arrest reaction with 20 mM EDTA when colour development is optimum.

## REVIEW QUESTIONS

1. What would have happened if you had eliminated the blocking step?

2. What would your developed membrane look like?

# 40

# MEASUREMENT OF PLANT CELL GROWTH

## OBJECTIVE

To measure plant cell growth.

## INTRODUCTION

Growth is defined as any process that results in a permanent increase in size or number. Plant growth is measured in many different ways like measuring the increase in root and shoot length, leaf area, stem girth, dry tissue weight and fruit volume. Cytologists measure the plant cell growth by measuring cell mass, packed cell volume and cell density.

A simple and accurate device for measuring cell density with the aid of microscope is the haemocytometer. The haemocytometer has a small glass-covered chamber designed to hold a specific volume of cell suspension culture. Researchers working with large cells prefer to use a Fuchs – Rosenthal haemocytometer, which has a deeper chamber (0.2 mm) than the conventional Neubauer haemocytometer (0.1 mm).

## MATERIALS

### Culture

*Pandorina morum* algae cell culture

### Media

L/10 Green algae medium

## OBSERVATION AND RESULT

| Days | Number of cells |
|---|---|
|  |  |
|  |  |

## Equipment

Mechanical counter
Light microscope

## Miscellaneous

Ethanol 95%
Lugol's solution
Fuchs–Rosenthol or Neubauer haemocytometer

## METHOD

1. Using a sterile 1-ml pipette inoculate the *Pandorina morum* culture in L/10 Green algae medium.
2. Shake the culture vessel for 15 seconds for even suspension of the cells. Calculate the initial cell density of the culture as follows.
3. Using a sterile 1-ml pipette transfer 1 ml of the culture to a sterile tube. Kill the cells by adding 1 ml of Lugols iodine to the culture tube.
4. Mix the contents of the tube by vortexing it thoroughly.
5. Transfer 1 ml of the diluted cell suspension to a second culture tube containing 9 ml of sterile distilled water. Label the dilution factor of this tube as 1/20.
6. Transfer 1 ml of the diluted cell suspension to a third culture tube which consists of 9 ml of sterile distilled water. Now the dilution factor of the cells is 1/20.
7. Using a clean 1-ml pipette, draw the diluted culture and load it in the haemocytometer.
8. Allow 1 minute for the cells to settle at the bottom of the haemocytometer. Focus the microscope so that the cells and grid markings are visible at 40X magnification.
9. Count the number of cells present in any one of the chambers. If the number of cells were less than 50 or more than 100, reload the chamber with cells from a different diluted culture.
10. After measuring the initial cell density incubate the culture at room temperature on an orbital shaker under cool white fluorescent lights.
11. Measure the cell density of the culture every day for two weeks and record your results.

12. Plot the number of cells against time in days on a graph sheet.


## REVIEW  QUESTIONS

1.  How will you measure the cell density?

2.  What is the difference between Fuchs–Rosenthol haemocytometer and
    Neubauer chamber?

# 41

# ESTABLISHING A PLANT CULTURE

## OBJECTIVE

To become familiar with plant tissue culture techniques.

## INTRODUCTION

The development of plant tissue culture technique has a significant impact on plant biotechnology research. Tissue culture involves the isolation of small pieces of living plant tissues called explants and their cultivation on nutrient media. The cells or tissues are obtained from any part of the plant like stem, root and leaf, which are encouraged to produce more cells in culture and to express their totipotency. The Murashige and Skoog (MS) based culture media commonly used for plant tissue culture has proved to be effective in cultivating both monocotyledons and dicotyledons. Besides from its use to study plant growth and development, plant tissue culture has many important economic uses including plant propagation, crop plant hybridisation and genetic engineering.

## MATERIALS

### Media

Sterile media  (MS medium)

### Miscellaneous

Plant material
10% Household bleach solution (prepared freshly)

## OBSERVATION AND RESULT

Within the first week explants can be expected to swell noticeably.

sterile water
70% Ethanol
95% Ethanol
Forceps
Scalpel

*Plant material*   A number of plants can be tested in this experiment. The best material for explants comes from plants raised indoors or in the green house because they seem to be cleaner. If you want to use vegetables from the grocery try to purchase as fresh as possible.


## METHOD

1. Select fresh-looking, healthy plant material. Cut the tissue into manageable pieces.
2. Carry the tissues into a sterile hood for disinfection.
3. Wash the explants in a beaker of distilled water, which consists of a few drops of detergent.
4. Using forceps, transfer the explants to a 70% ethanol solution for 5 minutes.
5. Again using the forceps transfer the explants to a solution of household bleach for 5–10 minutes. Tender leaf tissue should not be left in the bleach for more than 5 minutes.
6. After treatment with bleach, care should be taken to avoid re-contaminating it.
7. Rinse the explants in 3, five-minute changes of sterile distilled water.
8. Transfer the explant from the rinse water to the lid of a sterile petridish. For leaf tissue, cut the tissue into small squares, not more than 1 cm$^2$ using a sterile scalpel.
9. Cut the root or stem tissues into small cubes, 0.5–1.0 cm$^2$
10. Using sterile forceps, transfer the explants to the culture medium. Explants should be placed firmly in contact with the medium but it should not be buried. Incubate it.

# 42

# MICROPROPAGATION OF AFRICAN VIOLET

## OBJECTIVE

To demonstrate the principles of micropropagation and plant regeneration.

## INTRODUCTION

The purpose of this experiment is to demonstrate the principles of micropropagation and plant regeneration.

## MATERIALS

### Plant

Healthy African violet plant

### Media

Micropropagation media

### Miscellaneous

Sterile petri plates
Forceps and scalpel
NaOCl sterilisation solution
Ethanol 90% solution

## OBSERVATION AND RESULT

Examine the cultures regularly for the development of callus along the cut edges of the leaf segments.

## METHOD

1. Remove two clean young leaves (2–3 cm in diameter) from a healthy African violet plant.
2. Using a pair of forceps dip each leaf in ethanol solution for 30 seconds.
3. Transfer the leaves to a sterile petri plate and pour enough NaOCl solution to completely cover the leaves.
4. Treat the leaves for 10 minutes and using sterile forceps transfer the leaves to a new sterile petri plate and add enough sterile distilled water to cover the leaves.
5. Wash the leaves in sterile water for about three to four times and transfer the leaves to a sterile petri plate.
6. Using a sterile scalpel cut a square 1 × 1 cm out of the centre of the leaf. The square should contain the leaf midrib.
7. Cut the square leaf in half a cross the midrib. Using sterile forceps transfer the half square to the BF vessel containing micropropagation medium. Insert the lower half of the half square and replace the vessel cap.
8. Incubate the vessels at room temperature under a cool-white fluorescent light source.

# 43

# PREPARATION AND FUSION OF PROTOPLASTS

## OBJECTIVE

To become familiar with developing protoplasts.

## INTRODUCTION

A protoplast is a plant cell bound only by a plasmalemma and the methods that are available for removing the cell walls varies from species to species. The cell walls of some green algae can be removed by concentrating and purifying endogenous lytic enzymes produced during gametogenesis. To isolate protoplast from higher plant cells, researchers often use lytic enzymes obtained from plant-colonising fungi. Protoplasts from higher plants can also be isolated mechanically by breaking the cell wall under pressure with a French press or by slicing cells apart with a scalpel.

Enzymatic methods generate very large number of protoplasts in comparison to mechanical methods, but in some instances the lytic enzymes have deleterious effects on plant cell metabolism. However enzymatic methods are widely used by researchers.

## MATERIALS

### Plant material

Red onion

## OBSERVATION AND RESULTS

Fusion of protoplasts can be observed under the microscope.

*Enzyme solution*

| | |
|---|---|
| Cellulase | 1.5% |
| Macerase | 1.5% |
| Pectinase | 0.2% |
| Mannitol | 0.4 M |
| Glycine | 2.0% |

Filter the enzyme solution through a 0.45 µm filter before use.

*PEG Fusion buffer*

| | |
|---|---|
| Glucose | 0.3 M |
| $CaCl_2$ | 66 mM |
| PEG (Mol.wt. 1400) | 40% |

*Solution W10*

9 parts Solution A:

| | |
|---|---|
| Glucose | 0.4 M |
| $CaCl_2$ | 66 mM |
| Dimethyl sulphoxide | 10% |

1 part solution B:
0.3 M Glycine-OH buffer pH 10.5

## METHOD

1. Using a razor blade or scalpel, cut a small square of tissue from one of the layers of a red onion. Slice the tissue again to expose the cells that were buried in the centre of the square.
2. Place the half of the tissue containing both red and white cells to face down in a small pool of about 0.2–0.5 ml of enzyme solution. Cover the container in which the tissue is being digested to prevent the enzyme solution from evaporating. Examine the digestion periodically for the release of protoplasts.
3. After an hour remove the remaining bits of undigested tissue.
4. Remove 2 drops of protoplasts to a clean microscopic slide. Protoplasts are spherical structures. Check the slide under the microscope to see that you have both red and colourless protoplasts.
5. Add one drop of PEG fusion buffer to the protoplasts. Check the slide under the microscope and you can observe protoplasts clumping together.
6. After 15 minutes carefully remove the liquid from the slide and the protoplasts remain behind.

7.  Add 2 drops of solution W10 to the slide. Check the slide under the microscope periodically over the next 20 minutes during which fusion may begin to occur.

## REVIEW QUESTIONS

1.  Write notes on

    a. Protoplast.

    b. PEG.

# 44

# ISOLATION OF CHLOROPLASTS FROM SPINACH LEAVES

## OBJECTIVE

To become familiar with extracting chloroplasts.

## INTRODUCTION

Chloroplasts are the subcellular sites of photosynthesis, the process by which green plants, using energy from light, produce carbohydrate and oxygen from carbon dioxide and water. Under the microscope chloroplasts are recognised as bean-shaped, membrane-bound, green organelles.

## MATERIALS

### Plant sample

Fresh spinach leaves

### Equipment

Refrigerated preparative centrifuge
Haemacytometer and microscope

## OBSERVATION AND RESULT

Record the number of chloroplasts/ml__________.

## Reagents

1. Grinding solution
   - (a) 0.33 M Sorbitol
   - (b) 10 mM Sodium pyrophosphate
   - (c) 4 mM $MgCl_2$
   - (d) 2 mM Ascorbic acid
   - (e) Adjust pH to 6.5 with HCl
2. Suspension solution
   - (a) 0.33 M Sorbitol
   - (b) 2 mM EDTA
   - (c) 1 mM $MgCl_2$
   - (d) 50 mM HEPES
   - (e) Adjust pH to 7.6 with NaOH

## Miscellaneous

Chopping board and knife
Chilled mortar and pestle
Cheesecloth

## METHOD

1. Prepare an ice bath and pre-cool all the glasswares that are to be used.
2. Select several fresh spinach leaves and remove the large veins by tearing them loose from the leaves. Weigh out 4.0 grams of deveined leaf tissue.
3. Chop the tissue as fine as possible and add the tissue to an ice-cold mortar containing 15 ml of grinding solution and grind to a fine paste.
4. Filter the solution through double-layered cheesecloth into a beaker and squeeze the tissue pulp to recover all of the suspension.
5. Transfer the green suspension to a cold 50-ml centrifuge tube and centrifuge at 200 rpm for 1 minute at 4°C to pellet the unbroken cells and fragments.
6. Decant the supernatant into a clean centrifuge tube and recentrifuge at 1000 rpm for 7 minutes. The pellet formed during this

centrifugation contains chloroplasts. Decant and discard the supernatant.

7. Resuspend the chloroplast pellet in 5.0 ml of cold suspension solution or 0.035 M NaCl. Use a glass stirring rod to gently disrupt the packed pellet. This is the chloroplast suspension for use in subsequent procedures.

8. Enclose the tube in an aluminum foil and place it in an ice bucket.

9. Determine the number of chloroplasts/ml of suspension media using a haemocytometer.

## REVIEW QUESTIONS

1. What is the purpose of

   (a) Sorbitol

   (b) Sodium pyrophosphate

   (c) Ascorbic acid

   (d) EDTA

   (e) HEPES

# ISOLATION OF PLANT DNA

## OBJECTIVE

To become expert in extracting plant DNA.

## INTRODUCTION

Plant molecular biology is the study of the chemical and physical structure of biological macromolecules in plants. Recombinant DNA techniques are the powerful tools used today by plant molecular biologists. The ability to isolate, purify, splice and sequence DNA has made it possible to transfer genes from one organism to another, as well as to study the fundamental structure of genes and gene expression. Isolation of plant genomic DNA is a requirement for most genome characterisation, mapping and isolation of genes for genetic engineering. The degree of purity and quality of DNA required varies from one application to another. Good extraction procedure should yield DNA reasonably pure and intact. Since plant cells are having both cell wall and cell membrane, they must be removed one by one to release the cell contents. This can be achieved by combining two techniques, a technique to lyse gently the cells and solubilise the DNA followed by one of several basic enzymatic or chemical methods to remove contaminating proteins, RNAs and other macromolecules.

## MATERIALS

CTAB – extraction buffer (500 ml)
| | |
|---|---|
| Tris - HCl | 0.1 M |
| NaCl | 1.4 M |
| EDTA (pH 8.0) | 20.0 mM |

CTAB (Cetyl trimethyl ammonium bromide)     2.0% w/v
2 – Mercaptoethanol                                        0.1% w/v
Mercaptoethanol should be added only prior to use.

Liquid nitrogen
RNase A solution                                           10 mg/ml
Chloroform : Isoamylalcohol                          24 : 1
Ethanol                                                          70%

**TE buffer**
Tris – HCl (pH 8.0)                                        10 mM
EDTA (pH 8.0)                                             1 mM

Absolute alcohol, Isopropanol
NaCl                                                            3 M
Sodium acetate                                            3 M

## METHOD

1. Grind 2 grams of leaf tissue in liquid nitrogen using a mortar and pestle. Transfer the ground leaf tissues to a 50-ml centrifuge tube.
2. Add 15 ml of CTAB extraction buffer, preheated to 90°C and suspend thoroughly.
3. Incubate the suspension at 65°C for 30 minutes with occasional mixing and allow the mixture to cool to room temperature.
4. Add equal volume of chloroform : isoamyl alcohol (24 : 1) mixture and mix well to get an emulsion by inverting the tube several times for 15 minutes.
5. Centrifuge the mixture at 12,000 rpm for 5 minutes and separate the aqueous layer carefully.
6. Precipitate the chromosomal DNA by adding 0.7 volume of isopropanol, mix well and incubate at room temperature for 15 minutes.
7. Dissolve the DNA in minimal volume of TE buffer and incubate at 65°C for 10 minutes to accelerate the dissolution of DNA and inactivate residual DNase I.
8. Add 1/100 volume of 10 mg/ml RNase A solution. Mix and incubate at 37°C for 30 minutes.
9. Add equal volume of chloroform : isoamylalcohol (24 : 1) and centrifuge at 12,000 rpm for 10 minutes.
10. To the aqueous layer add 2 volumes of absolute alcohol and 1/10 volume of 3M sodium acetate.

11. Centrifuge at 10,000 rpm for 5 minutes and discard the supernatant.
12. Wash the precipitate with 70% ethanol and air-dry the pellet and dissolve it in a TE buffer.


## REVIEW QUESTIONS

1.  What is the purpose of

    (a)  CTAB


    (b)  Mercaptoethanol


2.  What are the methods available for extracting DNA from plants?


3.  What is the use of chloroform : isoamylalcohol?

# APPENDIX I

## LIST OF MEDIA

### Czapek dox agar

| | |
|---|---|
| Sodium nitrate | 2 gm |
| Dipotassium hydrogen orthophosphate | 1 g |
| Magnesium sulphate | 0.5 g |
| Potassium chloride | 0.5 g |
| Ferrous sulphate | 0.01 g |
| Sucrose | 0.01 g |
| Agar | 15 g |
| Distilled water | 1000 ml |

### Lactose broth

| | |
|---|---|
| Beef extract | 1 g |
| Proteose peptone | 10 g |
| Sodium chloride | 5 g |
| Lactose | 5 g |
| Phenol red | 0.018 g |
| Distilled water | 1,000 ml |
| Final pH | 7.4 |

### Minimal agar

| | |
|---|---|
| Sodium phosphate dibasic | 6 g |
| Potassium phosphate monobasic | 3 g |
| Sodium chloride | 0.5 g |
| Ammonium chloride | 1 g |
| Agar | 15 g |
| Distilled $H_2O$ | 1,000 ml |

After autoclaving, add 50 ml warmed, sterile 40% glucose, and swirl gently to mix.

## Micropropagation media

| | |
|---|---|
| Murashige and Skoog medium | 1.0 l |
| Indolebutyric acid (IBA) | 1.0 mg |
| Benzyl amino purine (BAP) | 3.0 mg |
| pH | 5.5 |

## Media L/10 green algae medium

| | |
|---|---|
| $Ca(NO_3)_2.4H_2O$ | 0.118 g |
| $MgSO_4.7H_2O$ | 0.0195 g |
| $K_2HPO_4$ | 0.0197 g |
| P – IV trace metal stock | 3.0 ml |
| Distilled water | 1,000 ml |
| pH | 7.0 |

P – IV Trace metal stock

| | |
|---|---|
| $Na_2EDTA$ | 0.750 g |
| $FeCl_3.6H_2O$ | 97.0 mg |
| $MnCl_2.4H_2O$ | 41.0 mg |
| $ZnCl_2$ | 5.0 mg |
| $CoCl_2.6H_2O$ | 2.0 mg |
| $Na_2MoO_4$ | 4.0 mg |

First dissolve the $Na_2EDTA$ in 500 ml of distilled water and then dissolve the remaining metal salts.

## Modified Czapek-mineral salt medium

| | |
|---|---|
| Sodium nitrate | 2 gm |
| Dipotassium hydrogen orthophosphate | 1 g |
| Magnesium sulphate | 0.5 g |
| Potassium chloride | 0.5 g |
| Ferrous sulphate | 0.01 g |
| Sucrose | 0.01 g |
| Cellulose | 10 g |
| Agar | 15 g |
| Distilled water | 1000 ml |

## Motility test agar

| | |
|---|---|
| Tryptose | 10 g |
| Sodium chloride | 5 g |
| Agar | 5 g |
| Distilled water | 1,000 ml |
| Final pH | 7.2 |

## MR–VP medium

| | |
|---|---|
| Peptone | 7 g |
| Dextrose | 5 g |
| Dipotassium phosphate | 5 g |
| Distilled water | 1,000 ml |
| Final pH | 6.9 |

## MS media

| | gm/l |
|---|---|
| 1.  Inorganic salts | |
| Macronutrients | |
| Ammonium nitrate | 1.65 |
| Potassium nitrate | 1.90 |
| Calcium chloride (Anhydrous) | 0.332 |
| Magnesium sulphate (Anhydrous) | 0.18 |
| Potassium phosphate | 0.17 |
| Micronutrients | |
| EDTA, Disodium salt | 0.0373 |
| Ferrous sulphate (heptahydrate) | 0.0278 |
| Manganese sulphate | 0.0169 |
| Zinc sulphate (heptahydrate) | 0.0086 |
| Boric acid | 0.0062 |
| Potassium iodide | 0.00083 |
| Sodium molybdate (dihydrate) | 0.00025 |
| Cobalt chloride (hexahydrate) | 0.000025 |
| Cupric sulphate (pentahydrate) | 0.000025 |
| 2.  Vitamins | |
| Myo –inositol | 0.1 |
| Thiamine hydrochloride | 0.0004 |
| 3.  Carbon source | |
| Sucrose | 30 |
| Agar | 7.5 |

## Mueller-Hinton agar

| | |
|---|---|
| Beef infusion | 300 g |
| Casamino acids | 17.5 g |
| Starch | 1.5 g |
| Agar | 17 g |
| Distilled water | 1,000 ml |
| Final pH | 7.3 |

## Nitrate broth

| | |
|---|---|
| Peptone | 5 g |
| Beef extract | 3 g |
| Potassium nitrate | 1 g |
| Distilled water | 1,000 ml |
| Final pH | 7.0 |

## Nutrient broth/agar

| | |
|---|---|
| Peptone | 5 gm |
| Sodium chloride | 5 gm |
| Beef extract | 3 gm |
| Agar | 15 gm |
| Distilled water | 1,000 ml |
| pH adjusted to | 7.2 |

## Oxidation-fermentation (O-F) glucose medium

| | |
|---|---|
| Glucose | 10 g |
| Tryptone | 2 g |
| Sodium chloride | 5 g |
| Dipotassium phosphate | 0.3 g |
| Bromothymol blue | 0.08 g |
| Agar | 2 g |
| Distilled water | 1,000 ml |
| Final pH | 6.8 |

## SIM medium

| | |
|---|---|
| Peptone | 30 g |
| Beef extract | 3 g |
| Peptonized iron | 0.2 g |
| Sodium thiosulfate | 0.02 g |
| Agar | 3 g |
| Distilled water | 1,000 ml |
| Final pH | −7.3 |

## Starch agar medium

| | |
|---|---|
| Starch | 20 g |
| Peptone | 5 g |
| Beef extract | 3 g |
| Agar | 15 g |
| Distilled water | 1,000 ml |
| pH | 7.0 |

## Tryptic soy agar/broth

| | |
|---|---|
| Trypticase | 15 g |
| Phytone | 5 g |
| Sodium chloride | 5 g |
| Agar | 15 g |
| Distilled water | 1,000 ml |

# APPENDIX II

### Methylene blue

| | |
|---|---|
| Methylene blue | 0.3 g |
| Ethyl alcohol | 30 ml |
| Water | 100 ml |

### Crystal violet

Solution A
| | |
|---|---|
| Crystal violet | 2 g |
| Ethanol | 20 ml |

Solution B
| | |
|---|---|
| Ammonium oxalate | 0.8 g |
| Water | 80 ml |

Mix solution A and B

### Grams iodine

| | |
|---|---|
| Potassium iodide | 2 g |
| Distilled water | 300 ml |
| Iodine | 1 g |

### Safranin

| | |
|---|---|
| Safranin | 0.25 g |
| Ethanol | 10 ml |
| Distilled water | 90 ml |

## Carbol fuchsin (Ziehl's)

Solution A
    Basic fuchsin (90% dye content)          0.3 g
    Ethyl alcohol (95%)          10 ml

Solution B
    Phenol crystals          5.0 g
    Distilled water          95 ml

Dissolve the solutes of each solution in the respective solvents indicated. Mix A and B and allow it to stand. Filter the final preparation and store it in bottles.

## Malachite green

    Malachite green          5.0 g
    Distilled water          100 ml

## Nigrosin

    Nigrosin          10 g
    Distilled water          100 ml
    Formalin          0.5 ml

Dissolve nigrosin in distilled water and immerse the mixture in boiling water bath for 30 minutes. Then add formalin and filter the solution twice through double filter paper.